SpringerBriefs in Applied Sciences and Technology

SpringerBriefs present concise summaries of cutting-edge research and practical applications across a wide spectrum of fields. Featuring compact volumes of 50 to 125 pages, the series covers a range of content from professional to academic.

Typical publications can be:

- A timely report of state-of-the art methods
- An introduction to or a manual for the application of mathematical or computer techniques
- A bridge between new research results, as published in journal articles
- A snapshot of a hot or emerging topic
- An in-depth case study
- A presentation of core concepts that students must understand in order to make independent contributions

SpringerBriefs are characterized by fast, global electronic dissemination, standard publishing contracts, standardized manuscript preparation and formatting guidelines, and expedited production schedules.

On the one hand, **SpringerBriefs in Applied Sciences and Technology** are devoted to the publication of fundamentals and applications within the different classical engineering disciplines as well as in interdisciplinary fields that recently emerged between these areas. On the other hand, as the boundary separating fundamental research and applied technology is more and more dissolving, this series is particularly open to trans-disciplinary topics between fundamental science and engineering.

Indexed by EI-Compendex, SCOPUS and Springerlink.

Jörg Kammermann · Igor Bolvashenkov ·
Staša Gejo · Andrey Brazhnikov ·
Alexander Rubinraut · Ilia Frenkel

Feasibility of Sustainable Electric Flights

In Air and Space

Jörg Kammermann ⓘ
Professorship of Energy Conversion
Technology
Technical University of Munich
Garching, Germany

Staša Gejo
Professorship of Energy Conversion
Technology
Technical University of Munich
Garching, Germany

Alexander Rubinraut
Expplanet Design Office
Munich, Bayern, Germany

Igor Bolvashenkov
Professorship of Energy Conversion
Technology
Technical University of Munich
Garching, Germany

Andrey Brazhnikov
Siberian Federal University (SibFU)
Krasnoyarsk, Russia

Ilia Frenkel
Shamoon College of Engineering
Beer Sheva, Israel

ISSN 2191-530X   ISSN 2191-5318   (electronic)
SpringerBriefs in Applied Sciences and Technology
ISBN 978-3-031-94784-1   ISBN 978-3-031-94785-8   (eBook)
https://doi.org/10.1007/978-3-031-94785-8

© The Author(s), under exclusive license to Springer Nature Switzerland AG 2025

This work is subject to copyright. All rights are solely and exclusively licensed by the Publisher, whether the whole or part of the material is concerned, specifically the rights of translation, reprinting, reuse of illustrations, recitation, broadcasting, reproduction on microfilms or in any other physical way, and transmission or information storage and retrieval, electronic adaptation, computer software, or by similar or dissimilar methodology now known or hereafter developed.
The use of general descriptive names, registered names, trademarks, service marks, etc. in this publication does not imply, even in the absence of a specific statement, that such names are exempt from the relevant protective laws and regulations and therefore free for general use.
The publisher, the authors and the editors are safe to assume that the advice and information in this book are believed to be true and accurate at the date of publication. Neither the publisher nor the authors or the editors give a warranty, expressed or implied, with respect to the material contained herein or for any errors or omissions that may have been made. The publisher remains neutral with regard to jurisdictional claims in published maps and institutional affiliations.

This Springer imprint is published by the registered company Springer Nature Switzerland AG
The registered company address is: Gewerbestrasse 11, 6330 Cham, Switzerland

If disposing of this product, please recycle the paper.

Preface

The challenge of sustainable electric propulsion has recently become an extremely important engineering direction, primarily because of the well-known advantages of electric traction drives. The most significant point of this direction is to develop the highly efficient and resilient traction electric traction drives, which will be the optimal choice for different types of vehicles used under specific operating conditions—in air and space.

The aim of this book is to present new solutions which make it feasible to realize effective and sustainable flights of aircraft and spacecraft with advanced electric propulsion systems.

The authors suggest that this book will be of considerable interest to researchers, practical engineers, and industrial managers who are involved in the addressing of issues related to the reliability-oriented design and the operation of fault-tolerant traction electric traction drives. In addition, it will be a helpful textbook for undergraduate and graduate courses in several departments, including electrical engineering, industrial engineering, mechanical engineering, and applied mathematics. This book is self-contained and does not require the reader to use other books or papers. There are four chapters in this book.

Chapter 1 focuses on the goals of air transport shift towards more-electric or all-electric airplanes. A major goal on the way towards the inclusion in commercial air traffic is high reliability. One of the experimental electric airplanes is NASA's X-57 "Maxwell", which consists of fourteen electric motors powered by a battery pack. The aim of this chapter is to assess the reliability of the proposed design of the X-57 by using the Lz-transform approach, as well as to propose several alternative designs to its electric drive train in order to use less vehicle mass on the motors and more on the battery pack, without sacrificing the original availability and expected performance, with a final goal to increase the sustainable flight range.

Chapter 2 presents traction electric motor designs developed for the use in the vehicles propulsion systems. Three basic design options for these motors were considered: an electric motor with a toroidal multiphase stator winding, an electric motor with a diamagnetic shield on the stator, and an electric motor with a multi-rod stator winding. The principles of construction of power supplies for motors with multi-rod stator winding have been developed. A comparison is made of post-emergency control strategies that can be used in the field of multiphase phase-pole controlled induction motor drives.

Chapter 3 is devoted to the project of a space electric train, named "asteroid-plane", for the regular delivery of astronauts onto the surface of the space bodies located in the asteroid belt. The asteroidplane is formed in the circumterrestrial orbit of the International Space Station from three electric rockets of the ER-7 type with electric rocket engines of the MARS type, in which working substance is nitrogen, and the source of electricity is a semiconductor solar battery. A new design solution for the asteroidplane in the presence of two takeoff and landing capsules will allow the regular delivering of 6 astronauts to the surface of any asteroid and searching for minerals there. The calculations have shown that the developed optimal trajectory of the asteroidplane flight between the orbit of the Earth and the orbit of the asteroid Psyche will allow to reach the asteroid Psyche in 55 days.

Chapter 4 is concerned with the project of electric space train "Titanplane" for astronauts regular delivery onto Titan—a satellite of the planet Saturn. At the first stage of the flight, with the help of a carrier rocket equipped with a chemical rocket engine, astronauts are delivered to the International Space Station. To deliver astronauts from Earth's orbit to Titan's orbit, a space train of six rockets equipped with superconducting electric and chemical rocket engines has been designed. For the next stage of the flight, a superconducting electric rocket engine has been proposed, equipped with three working chambers, in which the magnetic field is created by two oppositely connected flat electromagnets wound from superconducting wire. To supply the electric motor with electric power, an on-board power plant has been developed, consisting of a gas-phase nuclear reactor, a superconducting magneto-hydrodynamic alternator, and a cryoturbogenerator. The developed space train is capable of regularly delivering of six astronauts to any point on the surface of Titan in 65 days.

Garching, Germany	Jörg Kammermann
Garching, Germany	Igor Bolvashenkov
Garching, Germany	Staša Gejo
Krasnoyarsk, Russia	Andrey Brazhnikov
Munich, Germany	Alexander Rubinraut
Beer Sheva, Israel	Ilia Frenkel

Competing Interests The authors have no competing interests to declare that are relevant to the content of this manuscript.

Contents

1 **Comparison and Choice of Fault-Tolerant Traction Drive Topologies for a Full Electric Airplane** 1
   1.1 Introduction 1
   1.2 L_z-Transform Fundamentals 2
   1.3 NASA's X-57 "Maxwell" 4
      1.3.1 X-57 Specifications 4
      1.3.2 X-57 Cruise Motors and Propellers 5
      1.3.3 X-57 High-Lift Motors and Propellers 6
      1.3.4 X-57 Battery System 6
      1.3.5 X-57 Inverter 8
   1.4 Alternative Electric Propulsion System Units 9
      1.4.1 Considered Inverter Technologies 10
      1.4.2 Fault-Tolerant Electric Motors 13
   1.5 Proposed Alternative Topologies 14
      1.5.1 6PX Drive Topology 14
      1.5.2 9P6PX Drive Topology 16
      1.5.3 6P6 Drive Topology 17
      1.5.4 9P4 Drive Topology 18
      1.5.5 Results 19
   1.6 Conclusion 20
   References 21

2 **Fault-Tolerant Design and Control Strategies for Multiphase Traction Electric Drives** 25
   2.1 Introduction 25
   2.2 The Features of Controlled Multiphase Electric Drives 26
      2.2.1 Description of OPM 26
      2.2.2 Possibility to Decrease the Step of Speed Changing 30
   2.3 Design Implementation of the Law of H-Invariance 31
      2.3.1 On the Need to Fulfill the Law of H-Invariance 31
      2.3.2 Design 1 (D1) 33

		2.3.3 Design 2 (D2)	35

- 2.4 Design of the Drive System with Induction Motor 35
  - 2.4.1 Induction Motor of Design Version D3 35
  - 2.4.2 Alternative Version of Connection of D3 Design Motor to Energy Source 36
- 2.5 Further Development of AC Machine Concept with Stator Rod Winding 41
  - 2.5.1 Tubular-Rod Winding Filled with a Conductive Gaseous or Liquid Substance 41
  - 2.5.2 Windingless AC Motors and Generators 43
- 2.6 Additional Advantages of Multiphase Induction Motor Drives 46
  - 2.6.1 Strategies to Increase Reliability of Multiphase Induction Motor Drives 46
  - 2.6.2 Performance Evaluation of the Proposed Strategies 47
- 2.7 Conclusion 49
- References 49

3 Practical Application of Electric Propulsion Systems for the Regular Flights to the Asteroid Psyche 51
- 3.1 Introduction 51
- 3.2 The Concept of the Expedition to Psyche 52
- 3.3 The Structure of the Asteroidplane 61
- 3.4 Design of the Superconducting Electrorocket Motor MARS 63
- 3.5 Design of the Takeoff and Landing Capsule of the Asteroidplane ... 65
- 3.6 The Design of the Superconductor Bearing for Creation of an Artificial Gravity System in the Astronaut Cabin 66
- 3.7 The Basic Module for Creating Artificial Gravity During the Stay of the Expedition on the Surface of the Asteroid 67
- 3.8 Conclusion 68
- References 69

4 Feasibility of Electric Propulsion System for the Regular Flights onto Titan 71
- 4.1 Introduction 72
- 4.2 The Concept of Delivering Astronauts to Titan Using the Space Train "Titanplane" 73
- 4.3 Structure of the Titanplane 81
- 4.4 Space Locomotive of the Titanplane 81
- 4.5 Superconducting Electric Rocket Engine SERM-3K 83
- 4.6 On-Board Power Plant of the Titanplane 86
- 4.7 The Takeoff and Landing Capsule (TLC) of the Titanplane 88
- 4.8 Conclusion 89
- References 89

About the Authors

Dr.-Ing. Jörg Kammermann received his diploma (Dipl.-Ing.) in Electrical Engineering and Information Technology in 2011, as well as his doctoral degree (Dr.-Ing.) in Electrical and Computer Engineering in 2019, from Technical University of Munich (TUM) in Germany. From 2011 to 2016, he was a research associate and since 2016, he is an academic counselor with the Professorship of Energy Conversion Technology at TUM.

His research and teaching field includes the system analysis of electric vehicles based on application requirements, multiphase electric drives, and fault-tolerant electric drives for safety-critical applications.

Igor Bolvashenkov, Ph.D. is Senior Lecturer at the Institute of Energy Conversion Technology of Technical University of Munich (TUM), Munich, Germany. He obtained his M.Sc. (1981) and Ph.D. degrees (1989) in Electrical Engineering from Admiral Makarov State University of Maritime and Inland Shipping, Leningrad, USSR. From 1987 to 1993, he worked as an associate professor at the Murmansk State Technical University, Russia. Since 2004, he has worked at the Institute of Energy Conversion Technology at the Technical University of Munich (TUM), Munich, Germany.

He specializes in the development and simulation of electric propulsion system for ships, cars, trains, and aircrafts and comprehensive analysis of their efficiency, reliability, and fault tolerance. He has published five books, more than 160 scientific articles, chapters, and patents.

Staša Gejo, M.Sc. has obtained her bachelor degree in Electrical Engineering at the University of Maribor in Slovenia in 2019, specializing in power engineering. She graduated the Master of Science in Power Engineering program at the Technical University of Munich in 2023.

Her interests lie within the energy conversion technology and electric drives, their modeling, control, and reliability. Besides her academic pursuits, Staša is a professional sport climber, representing Serbia, a European Champion, World Games winner and a double World Championship bronze medalist.

Andrey Brazhnikov, Ph.D. is an associate professor at the Polytechnic Institute of the Siberian Federal University (SibFU), Krasnoyarsk, Russia. He obtained his B.Sc. degree (1982) in Automatics and Telemechanics from the Krasnoyarsk Polytechnic Institute, Krasnoyarsk, USSR, and he obtained his Ph.D. degree (1985) in Electrical Equipment from the Tomsk Polytechnic Institute, Tomsk, USSR. Since 2006, he has worked at the Siberian Federal University (SibFU), Krasnoyarsk, Russia.

He specializes in the development and simulation of multiphase AC electric drives, multiphase electromagnetic systems for stirring molten metals, and green energy generation systems (wind energy, sea wave's energy, etc.). He has published one book, more than 220 scientific articles, and patents.

Dr.-Ing. habil. Alexander Rubinraut, Design Office Expplanet, General Manager (Munich, Germany) graduated from The Moscow Energetic Institute in 1955, received a Ph.D. degree in 1967, habilitated in 1989, in Electrical Engineering and Energetics. For 20 years was the head of special problem laboratory at the Moscow Research Institute of Electrical Engineering, where electrical motors and generators, operating based on effect of superconductivity have been created. He is the author of the book *Cryogenic Electrical Motors* and two monographs. He has published more than 100 scientific works and invention patents.

Ilia Frenkel, Ph.D. is Researcher at the Center for Reliability and Risk Management, Shamoon College of Engineering (SCE), Beer Sheva, Israel. He obtained his M.Sc. degree in Applied Mathematics from Voronezh State University, Russia, and Ph.D. degree in Operational Research and Computer Science from the Institute of Economy, Ukrainian Academy of Science, Kiev, Ukraine. He has more than 45 years of academic and teaching experience at universities and institutions in Russia and Israel. From 2001 till 2018 he served as Senior Lecturer at the Industrial Engineering and Management Department and Chair of the Center for Reliability and Risk Management, Shamoon College of Engineering (SCE), Beer Sheva, Israel. After retiring in 2018 he is serving as Researcher at the Center for Reliability and Risk Management, Shamoon College of Engineering (SCE), Beer Sheva, Israel.

He specializes in applied statistics and reliability with application to preventive maintenance. He published five books and more than 100 scientific articles and chapters. He has edited 5 books and 12 special journal issues.

Chapter 1
Comparison and Choice of Fault-Tolerant Traction Drive Topologies for a Full Electric Airplane

Abstract As the goals of air transport shift towards more-electric or all-electric airplanes, different drive train configurations have been explored recently. A major goal on the way towards the inclusion in commercial air traffic is high reliability. One of the experimental electric airplanes is NASA's X-57 "Maxwell", which consists of fourteen electric motors powered by a battery pack. The aim of this paper is to assess the reliability of the proposed design of the X-57 by using the L_z-transform approach, as well as to propose several alternative designs to its electric drive train in order to use less vehicle mass on the motors and more on the battery pack, without sacrificing the original availability and expected performance, with a final goal to increase the flight range. The reliability analyses show that replacing the X-57's three-phase motors with six-phase ones greatly improves the availability of the electric drive train due to the use of fault-tolerant electric machines. Additionally, all the further proposed alternative designs have higher availability than the X-57. The alternatives with cascaded H-bridge (CHB) inverter topologies generally achieved higher availability values and higher expected performance than the six-pulse bridge (B6) variants. Finally, a distributed propulsion system with smaller take-off motors leads to a motor-mass advantage compared to more conventional drive train designs.

Keywords Aircraft reliability · L_z-transform · Comparative analysis · Electric airplane · Fault tolerance

1.1 Introduction

The European Union's climate action plan aims to reduce greenhouse gas emissions and become completely climate-neutral by 2050 [1]. One of the steps towards achieving net zero carbon emissions is the electrification of the transport sector, with a long-term goal to completely replace combustion engines with electrical engines. A shift towards the use of electrical drives has recently been seen in aviation, from hybrid-electric to all-electric projects. An all-electric aircraft is around two to three times more energy efficient compared to its fossil fuel counterpart [2]. In addition to improved efficiency and reduced carbon emissions, electrified aviation can also contribute to noise reduction. The main limitation of an all-electric aircraft is the flight range, as the current battery technology, in combination with the available volume and weight within the aircraft, allows for a total flight duration of only about 60 min [3, 4].

NASA is developing a battery-powered electric airplane X-57 "Maxwell" with an innovative distributed propulsion system, which allows for a reduction in wing area, while providing the same lift with less motor power [5, 6]. A similar distributed propulsion system is featured in the "ECO-150-300" [7] and the "DRAGON" [8] electric airplane models, as well as in the hybrid-electric plane design exploration in [9].

In order to become a part of commercial transportation, electric airplanes need to get certified by an airplane safety organization such as European Union Aviation Safety Agency (EASA) or Federal Aviation Administration (FAA). The standards for certifying electric airplanes are still in the making, however there are some preliminary documents or conditions in use (such as SC E-19 by EASA). High aircraft reliability is one of the main issues in its inclusion in the air transportation system. Different analytical methods can be used to assess system reliability, one of which is the L_z-transform method, suitable for complex multi-state systems. It was introduced in [10], as an extension of the widely used Universal Generating Function method, which can only be used for steady-state processes. In [11], the application of the L_z-transform to the dynamic reliability analysis of a multi-state system is explained. This method has been applied in reliability analyses of traction systems of ships [12] and helicopters [13], as well as manufacturing systems [14].

Reliability analysis can be a mathematically challenging task, especially in the case of complex multi-state systems. Modeling of continuous-time stochastic processes often requires a vast number of differential equations, which are usually computationally demanding and time-consuming to solve. The L_z-transform method allows for a quicker and simpler analysis of multi-state systems.

1.2 L_z-Transform Fundamentals

The L_z-transform of a discrete-state continuous-time Markov process $G(t)$ is mathematically represented as:

$$L_z\{G(t)\} = \sum_{i=1}^{k} p_i(t) z^{g_i}, \qquad (1.1)$$

where p_i is a probability that the process G is in a certain state i, out of k possible states, at a time instant t. Performance of the states is quantified with exponents g_i. As a process cannot be in more than one state at one time instant, the events are mutually exclusive and the sum of probabilities is equal to 1 [15]. In order to obtain

1.2 L_z-Transform Fundamentals

all the probabilities across the desired time frame, it is necessary to solve a system of differential equations given by:

$$\dot{p} = M^T p \qquad (1.2)$$

$$M^T = \begin{pmatrix} -\sum_{j \neq 1} x_{1j} & \mu_{21} & \cdots & \mu_{n1} \\ \lambda_{12} & -\sum_{j \neq 2} x_{2j} & \cdots & \mu_{n2} \\ \vdots & \vdots & \ddots & \vdots \\ \lambda_{1n} & \lambda_{2n} & \cdots & -\sum_{j \neq n} x_{nj} \end{pmatrix} \qquad (1.3)$$

where x is the respective matrix entry in the diagonal. The terms on the main diagonal of the matrix in Eq. (1.3) represent the sum of all "departing" transitions from each respective state of a Markov chain, denoted in the first index, whereas each of the remaining elements represent one single transition from and to a state defined in its index (see Fig. 1.1). To connect two elements in a series connection, the following Ushakov operator is applied, imposing a "min" function on the exponents of the two L_z-transforms:

$$L_z\{GF_s(t)\} = \Omega_{\text{ser}}(G(t), F(t)) = \sum_{i=1}^{k} \sum_{j=1}^{h} p_i(t) p_j(t) z^{\min(g_i, g_j)}. \qquad (1.4)$$

When it comes to a parallel connection of redundant elements, Ushakov operator imposes a "sum" function on the exponents:

$$L_z\{GF_p(t)\} = \Omega_{\text{par}}(G(t), F(t)) = \sum_{i=1}^{k} \sum_{j=1}^{h} p_i(t) p_j(t) z^{g_i + g_j}. \qquad (1.5)$$

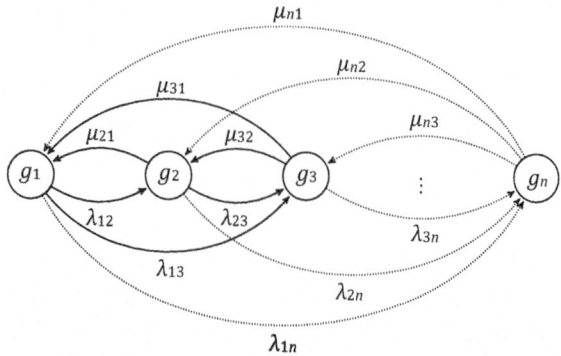

Fig. 1.1 Markov chain representation for an n-state process

The final L_z-transform of a complete multi-state system can be directly used for calculating important parameters which give an insight about the reliability of the system [15].

Availability is calculated as the sum of all probabilities of the exponents that fulfill the criterion, i.e. are greater than or equal to the defined ω, shown in Eq. (1.6). Availability is a quantity used to describe the ability of a system to function without failures, as well as the ability of the external environment to repair the system in case of failure, to bring it back to the healthy state [16]. Furthermore, it is the ability of the system to function after a failure, due to its fault tolerance.

$$A(t, \omega) = \sum_{g_i \geq \omega} p_i(t). \tag{1.6}$$

Expected performance is a sum of the products of probabilities and the exponents:

$$E(t) = \sum_{i=1}^{k} p_i(t) g_i. \tag{1.7}$$

1.3 NASA's X-57 "Maxwell"

In recent years, NASA has been developing an experimental electric aircraft called X-57 Maxwell, with a distributed propulsion system. It is based on Tecnam P2006T plane, whose internal combustion motors have been removed in order to test the electric propulsion systems through the four defined project stages. The idea behind the distributed propulsion system lies in increasing aerodynamic lift by the use of thrust from light motors spread across the wings, therefore allowing for a reduction of the size of wings for a more efficient cruise [17].

1.3.1 X-57 Specifications

The modification IV of the X-57 airplane utilizes a distributed propulsion system which consists of:

- two 47 kWh battery packs,
- two 60 kW cruise motors and propellers, and
- twelve 10.5 kW high lift motors (HLM) and foldable propellers [17].

All fourteen motors are used at takeoff and climb, and only the two cruise motors are used in cruise mode, when the propellers of HLM fold to reduce drag. More on

1.3 NASA's X-57 "Maxwell"

specific flight phases can be found in [4, 18]. The duration of the flight is 3400 s or approximately 57 min. Two flight phases/demands are considered within this chapter:

- the Roll phase, which demands 240 kW for 10 s and
- the Climb1 phase, which demands 210 kW for 90 s,

as the other phases from [18] represent a demand that is too low for a representative comparison of the availability functions.

The battery weighs more than the rest of the propulsion system with a total mass of 390 kg. Mass of cruise motors is reportedly 22.2 kg each [19], whereas one HLM has a motor mass of 2.34 kg [20]. The total mass of all motors is 72.5 kg, which will be important in comparison with alternative drive train topologies. The total aircraft mass is 1360 kg [17]. It is important to note that this aircraft is still in the process of development, which means that documents and specifications have changed throughout the process, and could still be subject to future changes.

The propulsion system was simplified to the following main components: battery packs, inverters, motors, and propellers, which will be covered with respect to reliability in the following sections. A reliability analysis was performed by means of the L_z-transform method. The calculated reliability indices will be used for comparison with proposed alternative propulsion system designs.

1.3.2 X-57 Cruise Motors and Propellers

The two cruise motors at the ends of the wings are 60 kW three-phase PMSM air-cooled out-runner motors, produced by Joby Aviation [19]. Cruise propellers in use are of variable-pitch, consisting of three blades [21].

A three-phase motor is considered as a two-state Markov process, named "M". Markov chain representation for two-state elements is shown in Fig. 1.2, where any phase failure results in a total failure of the motor, since it requires all three phases to run. The performance of the healthy state g, as well as the transition rates depend on the element.

As it operates at its nominal power of 60 kW, the Markov process states for the cruise motor are 60 and 0. This element is represented using L_z-transform as shown in Eq. (1.8). Failure and repair rates for this process are chosen from [22] for a three-phase machine: $\lambda_M = 0.09$ y^{-1} and $\mu_M = 113$ y^{-1}

$$L_z\{M(t)\} = p_1^M(t)z^{60} + p_2^M(t)z^0. \tag{1.8}$$

Fig. 1.2 Markov chain representation for two-state elements

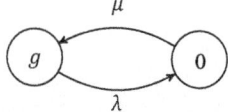

The propeller is also considered as a two-state process, named "P", with states 60 and 0, represented with L_z-transform in (1.9). The failure rate is chosen from the propeller data in NASA's General Aviation Aircraft Reliability Study [23] to be $\lambda_P = 0.066$ y^{-1}. The repair rate for a variable-pitch propeller is $\mu_P = 95$ y^{-1}, sourced from [24]. This source used a very similar failure rate of 0.06 y^{-1}, comparably to the aforementioned chosen one.

$$L_z\{P(t)\} = p_1^P(t)z^{60} + p_2^P(t)z^0 \tag{1.9}$$

1.3.3 X-57 High-Lift Motors and Propellers

X-57 is special because of its distributed propulsion system, which uses the small motors placed across the wings to increase lift and, therefore, enable the use of narrower wings [20]. Such design is meant to also increase efficiency of operation, providing more degrees of freedom in control, redundancy, and noise reduction [25].

X-57 Maxwell uses twelve 10.5 kW motors only during takeoff and landing. The propellers are foldable to reduce drag when not in use [20]. Both HLM and its propeller are treated as two-state processes "m" and "p", respectively, with performance values of 10.5 and 0. The L_z-transform of these two elements is shown in Eqs. (1.10) and (1.11).

$$L_z\{m(t)\} = p_1^m(t)z^{10.5} + p_2^m(t)z^0 \tag{1.10}$$

$$L_z\{p(t)\} = p_1^p(t)z^{10.5} + p_2^p(t)z^0 \tag{1.11}$$

The failure and repair rates for these two components are the same as for the cruise motor and cruise propeller, summarized in Eqs. (1.12) and (1.13).

$$\lambda_M = \lambda_m = 0.09 \text{ y}^{-1}, \mu_M = \mu_m = 113 \text{ y}^{-1} \tag{1.12}$$

$$\lambda_P = \lambda_p = 0.066 \text{ y}^{-1}, \mu_P = \mu_p = 95 \text{ y}^{-1} \tag{1.13}$$

1.3.4 X-57 Battery System

The battery system is made of two parallel battery packs, each consisting of eight modules. One module has 320 single NCA 18650-30Q cells, connected in a 20p16s configuration. Total battery system useful capacity is 47 kWh, it weighs 390 kg and the nominal voltage is 461 V. NASA's top-level battery requirements [4] are to

1.3 NASA's X-57 "Maxwell"

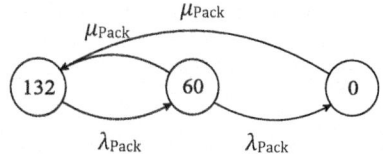

Fig. 1.3 Markov chain representation for a single battery pack

provide the maximum of 132 kW for 45 s, 74 kW for 3 min, and continuous power of 60 kW.

According to NASA requirements, a Markov model is defined per battery pack. Since the 74 kW per pack are insufficient to power neither Roll nor Climb1 phase, a three-state model of "Pack" is considered, the performance values of states being 132, 60, and 0, as graphically shown in Fig. 1.3 and mathematically in Eq. (1.14).

$$L_z\{\text{Pack}(t)\} = p_1^{\text{Pack}}(t)z^{132} + p_2^{\text{Pack}}(t)z^{60} + p_3^{\text{Pack}}(t)z^0 \quad (1.14)$$

Choosing the failure rates was a challenging task, as there is little data on the MTTF of battery packs in electric vehicle and aircraft applications. An option could be to use the warranty period as MTTF, which differs between the manufacturing companies. Studies on reliability of electric airplanes have used failure rates of 0.08 y^{-1} in [24], 150 FIT or 0.0013 y^{-1} in [26], 0.438 y^{-1} in [13]. Furthermore, $2.4 \cdot 10^{-6}$ h^{-1} or 0.021 y^{-1} was calculated in [27] and the value gets even as low as 0.0006 y^{-1} in [28].

On the other hand, there is extensive research available on the effect of cycle depth, temperature, discharge voltage, and current on capacity fade, and therefore aging. One must also take into account calendar aging, storage conditions, and charging. Data sheet of the Samsung 18650-30Q predicts 250 or 300 cycles to reach the end-of-life at maximum current discharge (15A). The authors of [29] report an average cycle life of 517, [30] reported 1024, and the average cycles to failure in [31] for two different types of cells were 587 and 724 under 1C discharge. If the value of 1024 is considered, at 1C discharge, it would take 2048 h^{-1} to failure (with 2 h in total to charge and discharge). This means that a single cell would fail around 4.3 times a year. From experience, it is known that our smartphone or laptop batteries last at least 1.5 years. Tesla provides a warranty of 8 years, Nissan of 2 years, and even Samsung's own single-cell warranty is 18 months. It is clearly noticeable that a connection between cycle life experiments and real-life capacity fade is not distinguishable.

However, a dissertation on aging of EV battery by Keil [32] provides more insight on specific parameters and their effect on capacity fade. Cycle depth was a dominant factor in cycling life, leading to a distinctly faster capacity fade compared to the effect of temperature, voltage, dynamics, etc. A long-term discharge test showed that a 61% depth of discharge caused the capacity fade to 80% after 850 days, or

2.33 years. A test to compare dynamic driving mode and constant current mode, with the same average current, gave similar capacity fade results.

In relation to the load profile of X-57, high power modes Roll and Climb1 discharge 13% of the total capacity in 100 s (other flight phases are not taken into account), which is considered a low depth of discharge. Therefore, double the power output will be assumed not to have a significant impact on the cycle fade and, therefore, will be assigned the same approximate time to failure as in the rest of the phases.

To conclude, the failure rates of the battery used in this analysis will be based on the period of 850 days to failure, resulting in $\lambda_{12}^{Pack} = \lambda_{23}^{Pack} = 0.429 \text{ y}^{-1}$. The repair rate was chosen as $\mu_{21}^{Pack} = \mu_{31}^{Pack} = 461 \text{ y}^{-1}$ from the average repair rate of batteries, found in [33].

When it comes to the RBD of X-57, the representation of the battery system is a simplification of the real scheme since both batteries provide half the power to each of the cruise motors. As this kind of configuration does not allow for a series–parallel reliability block diagram (RBD), a parallel connection of the two packs powering the whole system was chosen as the best approximation. A full traction scheme is available in [5, 34].

1.3.5 X-57 Inverter

It is briefly mentioned in [5] that each of the two power trains, consisting of three half-bridge modules, provides half the torque in a cruise motor, with the use of Cree's CAS300M12BM2 half-bridge chips. No further inverter information was found in NASA's literature.

Since [35] used the same module in reliability calculations, the adapted failure rate of $\lambda_{INV} = \lambda_{inv} = 0.0316 \text{ y}^{-1}$ will be used in the later reliability calculations. The value was adapted with the use of provided equations for three CAS300M12BM2 per module instead of the six considered in the study.

The repair rate was chosen from [24], with the value of $\mu_{INV} = \mu_{inv} = 584 \text{ y}^{-1}$. As two inverters are used in a redundant structure to power the cruise motors, they will be connected in parallel. Additionally, it is assumed that only one such inverter powers the HLM.

Reliability block diagram of X-57 is presented in Fig. 1.4. Capital letters are for the cruise motor branch elements, small letters for the HLM branches. There are twelve HLM branches and two cruise branches.

The inverters for a cruise motor ("INV") and the inverters for HLM ("inv") are two-state processes, as it is described in Fig. 1.2. Due to the redundant structure, there will be two parallel "INV" elements connected to the "M", each having half the branch performance − 30 and 0. The HLM inverter "inv" will have states 10.5 and 0. See the L_z-transforms of the two inverter elements in (1.15) and (1.16).

$$L_z\{INV(t)\} = p_1^{INV}(t)z^{30} + p_2^{INV}(t)z^0 \qquad (1.15)$$

1.4 Alternative Electric Propulsion System Units

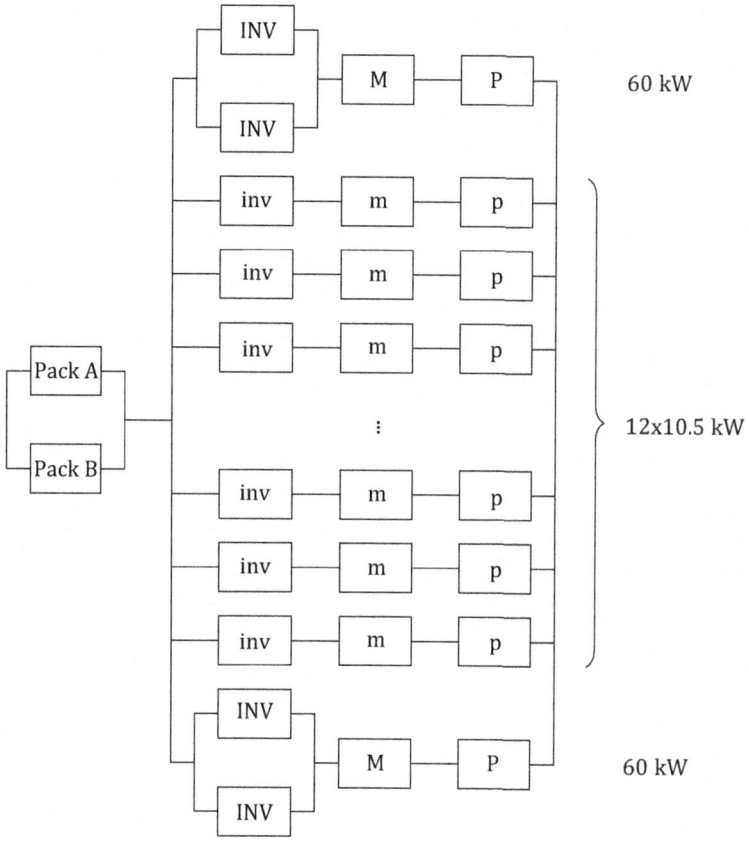

Fig. 1.4 Reliability block diagram of X-57

$$L_z\{\text{inv}(t)\} = p_1^{\text{inv}}(t)z^{10.5} + p_2^{\text{inv}}(t)z^0 \qquad (1.16)$$

1.4 Alternative Electric Propulsion System Units

Several alternative designs of the propulsion system are proposed with a lower total number of motors with more fault-tolerant and redundant elements, while meeting the power requirements. An additional goal was to achieve a lower total mass of the motors than the X-57. This way, the difference in mass could be used for the battery system, which could increase the flight range.

Two different inverter technologies are proposed and discussed, as well as the use of fault-tolerant machines. These are combined in different drive train configurations, whose reliability is analyzed and compared among each other, as well as with the X-57.

1.4.1 Considered Inverter Technologies

A simple half-bridge inverter used in X-57 has the advantage of a low number of semiconductors, taking up less volume and mass, and being controlled in a simpler manner compared to more complex inverter topologies. However, it lacks fault tolerance, which is very important for the availability of an electric aircraft drive train. A cascaded H-bridge (CHB) inverter is a multilevel inverter which offers a large scale of fault tolerance, depending on the number of levels, i.e. the number of modules used. The authors of [36] compared a CHB to a conventional six-pulse (B6) bridge in an electric helicopter and found a significantly smaller power ripple in the CHB topology during a single-phase failure, as well as better fault tolerance, especially for multiphase machines.

When it comes to failure rates, a simple and common strategy is to calculate the values relative to the number of components, based on the values from the Military Handbook [37]. In some cases, the base rates are adapted with experimental test results. A consequence of using this approach is that inverters with a lower number of components tend to have lower failure rates compared to the more branched or modular topologies. However, simple structures lack fault tolerance. Two different inverter topologies will be used in the analysis to compare the approaches: B6-bridge and 5-level CHB. Each variant will be assessed with two different base failure rates: one theoretical, based on the number of components and their individual failure rates, and one more practical, based on the results of available experimental studies.

1.4.1.1 B6-Bridge

This configuration includes one B6-bridge per three phases of a motor, allowing a multiphase machine to operate as a multiple of three phases, i.e. a nine-phase machine would operate as a triple three-phase machine. The total number of switches of this configuration is $2m$, with m denoting the machine's number of phases, as there are six switches per three phases in a B6-bridge. The chosen approach includes with one DC link for all the modules in the parallel connection, similarly as in the case of the X-57 cruise motor.

This topology is represented in an RBD as a parallel connection of $m/3$ branches, as depicted in Fig. 1.5. Each inverter element, named "B6", has two states, since a failure of one switch in a module leads to a failure in the whole module. The performance values are defined as P_{Mm} and 0, as each branch needs to be capable of carrying the full motor power P_{Mm} in case of failure of any of the other branches.

1.4 Alternative Electric Propulsion System Units

Fig. 1.5 Partial RBD example of the B6 topology with a nine-phase machine

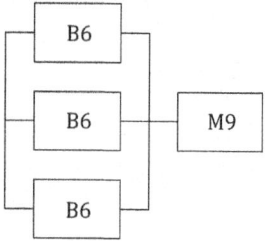

$$L_z\{B6(t)\} = p_1^{B6}(t)z^{P_{Mm}} + p_2^{B6}(t)z^0 \tag{1.17}$$

The practical failure rate was calculated by the formulae and data provided in [35] for one capacitor and three half-bridge modules, equals to $\lambda^{B6} = 0.0316$ y^{-1}. The same value was used for the X-57 cruise inverter (see Sect. 1.3.5). Similarly, the repair rate is $\mu^{B6} = 584$ y^{-1} from [24].

1.4.1.2 5-Level CHB

A 5-level CHB consists of two H-bridge modules per phase. Each module has its own separate source of energy. For that purpose, the original X-57 battery system requires reallocation and division into smaller units. Instead of using the two parallel battery packs, as in the B6 topology, each CHB module is powered from one small battery block. With this approach, the total battery system is physically split into $2N$ independent blocks, with N representing the total number of motor phases in the system.

If one of the two inverter modules fail with the other still functioning, that phase stays functional too, with an overload on the remaining inverter module. All phases are operated independently, but if one phase fails, the fault tolerance of multiphase machines allows for continued operation with one phase less. To model this behavior as precisely as possible with Markov models, each battery–inverter phase branch will be represented with a three-state "CHB" element in series with a three-state battery block ("BB") element. This is a minor simplification, as there are two modules of the inverter per phase and two sources. However, using three states for one element simulates the abovementioned failure degradation within the phase. If both modules fail, the third (failed) state is reached, meaning that the whole phase has failed.

The "BB" and "CHB" branches which power one motor are all connected in parallel with each other, then continue in series with the multiphase machine (see Fig. 1.6). Generally, one CHB module would technically have states P_{Mm}/m and 0, as it needs to be able to power one phase individually, if the other module fails. However, when the two modules are evaluated as one reliability element, as previously elaborated, the three states are $2P_{Mm}/m$, P_{Mm}/m and 0, shown in Eqs. (1.18) and (1.19).

Fig. 1.6 Partial RBD for CHB implementation for an m-phase machine

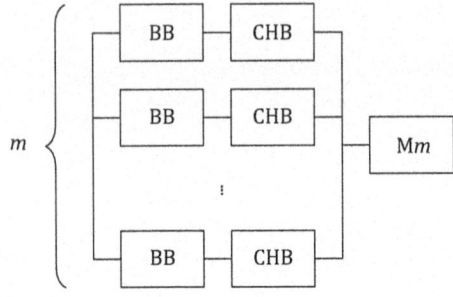

$$L_z\{\text{CHB}(t)\} = p_1^{\text{CHB}}(t)z^{g_1\text{CHB}} + p_2^{\text{CHB}}(t)z^{g_2\text{CHB}} + p_3^{\text{CHB}}(t)z^0 \quad (1.18)$$

$$g_1^{\text{CHB}} = 2g_2^{\text{CHB}} = 2\frac{P_{Mm}}{m} \quad (1.19)$$

Similarly, the "BB" element is defined with three states as well, to fully power one phase. Its L_z-transform is presented in Eq. (1.20). Newly calculated states of these battery blocks depend on the fraction of the total power, which a branch provides, and on the number of phases of the motor in the branch. A general formula on how to define new states of "BB" is provided in Eq. (1.21). As this formula is related to the original states of X-57 battery pack, the reader can revise the values in Eq. (1.14).

$$L_z\{\text{BB}(t)\} = p_1^{\text{BB}}(t)z^{g_1\text{BB}} + p_2^{\text{BB}}(t)z^{g_2\text{BB}} + p_3^{\text{BB}}(t)z^0 \quad (1.20)$$

$$g_i^{\text{BB}} = \frac{P_{Mm}}{P_{\text{total}}} \cdot \frac{2g_i^{\text{Pack}}}{m} \quad (1.21)$$

The failure rate for one "CHB" reliability element is calculated by using the data and equations of [38] for two capacitors, eight switches, and eight diodes, resulting in $\lambda_{12}^{\text{CHB}} = 0.206$ y^{-1}. This value is used for the first transition. However, due to the increased current as a consequence of one module failure, the functional one gets overloaded, which shortens its remaining lifetime. The current becomes two times higher, and the heating losses, therefore, increase four times, according to Joule's law. With these consequences in mind, the new failure rate for the second transition is calculated as four times the first transition, $\lambda_{23}^{\text{CHB}} = 0.824$ y^{-1}.

For a focus on failure comparability of the two topologies, the repair rate of one "CHB" element is equally defined as for the B6 inverter, i.e. $\mu_{21}^{\text{CHB}} = \mu^{B6} = 584$ y^{-1}, which covers just the repair after the first failure. The repair after the second failure, which involves two modules, meaning that the repair rate is half the first value: $\mu_{31}^{\text{CHB}} = 292$ y^{-1}.

1.4.2 Fault-Tolerant Electric Motors

Multiphase electric motors are considered fault-tolerant as a failure in one phase minimally obstructs the function of the machine. As shown in [22], higher number of phases means an increase in availability, with immediate repairs done after every phase failure. However, a failure of a phase may lead to a reduced amount of torque. Besides fault tolerance, multiphase machines have reduced harmonic content of the air gap, which leads to reduced torque ripple [39]. They require lower DC-link capacitors and have higher efficiency compared to conventional three-phase motors. Additionally, development of inverter technologies and modulation techniques allowed for more research to be conducted and more frequent application of the multiphase machines in both industry and academia [40].

The two motor choices for the alternative drive train configurations are six- and nine-phase machines. As previously mentioned, these can run each phase independently or in groups of three, acting as a multiple of a three-phase machine. Different inverter topologies of multiphase machines are discussed in [36, 39, 41, 42].

The authors of [43] proposed a methodology for defining the states of a multiphase machine as a Markov process. This paper covers detailed instructions on determining the number and performance values of states, depending on the performance demands. With the use of this method, a condition that a minimum of half of the machine power needs to be provided after an occurred failure. A six-phase machine will have five states:

$$g^{M6} = \{100\%, 83\%, 66\%, 50\%, 0\} P_{M6},$$

whereas a nine-phase machine will have six states:

$$g^{M9} = \{100\%, 88\%, 77\%, 66\%, 55\%, 0\} P_{M9}$$

The L_z-transform of the "M6" and "M9" elements is shown in Eqs. (1.22) and (1.23), respectively. Time dependency is considered in the L_z-transforms but will be omitted in writing from this point on for better readability.

$$L_z\{M6\} = p_1^{M6} z^{g_1 M6} + p_2^{M6} z^{g_2 M6} + p_3^{M6} z^{g_3 M6} + p_4^{M6} z^{g_4 M6} + p_5^{M6} z^0 \quad (1.22)$$

$$L_z\{M9\} = p_1^{M9} z^{g_1 M9} + p_2^{M9} z^{g_2 M9} + p_3^{M9} z^{g_3 M9} + p_4^{M9} z^{g_4 M9} + p_5^{M9} z^{g_5 M9} + p_6^{M9} z^0 \quad (1.23)$$

Failure and repair rates are taken from [22], as it was previously done for a three-phase machine in the X-57 reliability analysis. For a six-phase machine $\lambda^{M6} = 0.15 \text{ y}^{-1}$, $\mu^{M6} = 107 \text{ y}^{-1}$ and for a nine-phase machine $\lambda^{M9} = 0.21 \text{ y}^{-1}$, $\mu^{M9} = 97 \text{ y}^{-1}$.

1.5 Proposed Alternative Topologies

This section introduces different alternatives to the X-57 design. The idea is to use a lower number of parallel drive train branches, with more fault-tolerant structures and topologies. The drive topologies are named in a way to provide a hint on the structure of the topology, for example "X" meaning a distributed propulsion system (similar to X-57) and the numbers combined with a "P" refer to the motors' phase numbers. The 6PX copies the design of X-57, but replaces three-phase with six-phase motors, while all remaining component definitions stay unchanged. Other alternatives, being 9P6PX, 6P6, and 9P4, will use four to eight electric motors, distributed across the wings. Aerodynamic performance is neglected in this analysis.

A crucial assumption needed to be made ahead of the analysis related to the specific power of motors. The reported active mass of Joby's 60 kW cruise motor used in X-57 is 22.2 kg, with a specific power of 2.7 kW/kg. HLM's active mass of 2.34 kg and 10.5 kW give a specific power of approximately 4.5 kW/kg. El Refaie and Osama [44] overviewed different electric motors in aero and land vehicle applications, where the motors between 10 and 100 kW have power densities in the range of 2.4 and 5 kW/kg. A smaller motor of 10 kW was designed in [45] using lightweight stator construction and HiperCo50 magnet to result in 6 kW/kg specific power with only 1.67 kg. A 60 kW five-phase IPMSM was reported in [46] achieving 3.12 kW/kg. A six-phase electric aircraft motor was assessed in [47], resulting in 10.6 kW/kg for active parts (6.4 kW/kg considering total weight). Additionally, Da Rosa et al. [48] and Fleitas et al. [49] showed that a multiphase machine may not differ in mass from the three-phase machine and can even have a lower value, due to the lower volume of conductive material required.

In order to approximate the total motor mass in different alternative configurations, the specific power of motors up to 30 kW is taken as 4.5 kW/kg, otherwise, for powers higher than 30 kW, it is taken as 3.12 kW/kg. This limit was taken as an assumption based on overview of the air-cooled motors between 10 and 100 kW in [44]. Table 1.1 gathers all the failure and repair rates used for solving the systems of differential equations of Markov processes—elements in the drive train of X-57 and in the alternative drive trains.

1.5.1 6PX Drive Topology

This alternative uses the same design (position and mass of motors) as the original X-57 (see Fig. 1.7), however, includes six-phase instead of three-phase motors. An additional adaptation is the use of two B6 inverters per motor, applying also to the HLM, in order to properly run every six-phase motor.

The difference to X-57 in L_z-transform is found in the B6 inverters and motors. The new expressions are given in Eqs. (1.24)–(1.27), where the "M6" stands for the new cruise motor, "m6" for the new take-off motor, "$B6_M$" is cruise inverter, and

1.5 Proposed Alternative Topologies

Table 1.1 Elements of X-57 (upper block) and of the alternative drive train configurations (lower block), their number of states, and chosen failure and repair rates in year^{-1}

Element	Number of states	Failure rate	Repair rate
Pack	3	0.429	461
INV	2	0.0316	584
inv	2	0.3	584
M	2	0.09	113
m	2	0.09	113
P	2	0.066	95
p	2	0.066	95
B6	2	0.0316	584
CHB	3	0.206, 0.824	584, 292
M6	5	0.15	107
M9	6	0.21	97
P	2	0.066	95
BB	3	0.429	461

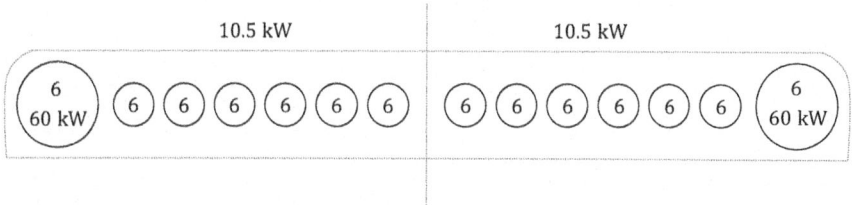

Fig. 1.7 6PX representation across the wings. The power distribution is the same as in X-57 only with six-phase motors

"B6$_m$" take-off branch inverter.

$$L_z\{B6_M\} = p_1^{B6_M} z^{60} + p_2^{B6_M} z^0 \quad (1.24)$$

$$L_z\{B6_m\} = p_1^{B6_m} z^{10.5} + p_2^{B6_m} z^0 \quad (1.25)$$

$$L_z\{M6\} = p_1^{M6} z^{60} + p_2^{M6} z^{50} + p_3^{M6} z^{40} + p_4^{M6} z^{30} + p_5^{M6} z^0 \quad (1.26)$$

$$L_z\{m6\} = p_1^{m6} z^{10.5} + p_2^{m6} z^{8.72} + p_3^{m6} z^{6.9} + p_4^{m6} z^{5.3} + p_5^{m6} z^0 \quad (1.27)$$

The comparison of results in Roll and Climb1 phase shows higher availability of 6PX alternative within both demands (Fig. 1.8a), as well as better expected performance than X-57 (Fig. 1.8b).

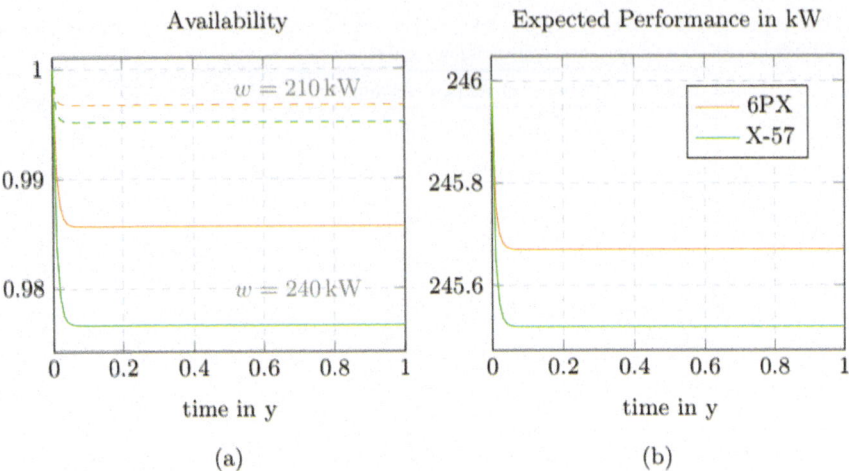

Fig. 1.8 Availability comparison of 6PX and X-57 (**a**) for $w = 210$ kW (dashed plots) and $w = 240$ kW (solid plots). Expected performance comparison in kW in (**b**)

1.5.2 9P6PX Drive Topology

A similar distributed propulsion system is used in this alternative, utilizing two nine-phase 60 kW cruise motors. There are six instead of twelve take-off motors of 21 kW each, all six-phase (see Fig. 1.9). With the previously mentioned specific power assumptions, the total motor mass of this alternative is estimated at 66.5 kg.

The L_z-transform for the 9P6PX of nine-phase, six-phase motors, and propellers are listed in Eqs. (1.28)–(1.31), the battery packs and B6 inverters are given in Eqs. (1.32)–(1.34), and lastly CHB with "BB" in Eqs. (1.35)–(1.38). The "Pack" equation for this and all the remaining alternatives is identical to the Eq. (1.14) and will not be repeated in the upcoming sections.

$$L_z\{M9_{60}\} = p_1^{M9_{60}} z^{60} + p_2^{M9_{60}} z^{52.8} + p_3^{M9_{60}} z^{46.2} + p_4^{M9_{60}} z^{39.6} + p_5^{M9_{60}} z^{33} + p_6^{M9_{60}} z^{0} \tag{1.28}$$

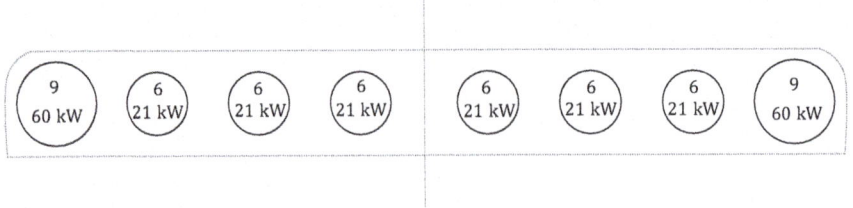

Fig. 1.9 9P6PX representation across the wings. It uses a distributed propulsion system with less motors than X-57

1.5 Proposed Alternative Topologies

$$L_z\{M6_{21}\} = p_1^{M6_{21}} z^{21} + p_2^{M6_{21}} z^{17.4} + p_3^{M6_{21}} z^{13.9} + p_4^{M6_{21}} z^{10.5} + p_5^{M6_{21}} z^0 \quad (1.29)$$

$$L_z\{P_{60}\} = p_1^{P_{60}} z^{60} + p_2^{P_{60}} z^0 \quad (1.30)$$

$$L_z\{P_{21}\} = p_1^{P_{21}} z^{21} + p_2^{P_{21}} z^0 \quad (1.31)$$

The B6 inverter topology is implemented in three branches for the nine-phase cruise motors and two branches for the rest of the smaller six-phase motors (refer to the RBD in Fig. 1.5).

$$L_z\{\text{Pack}\} = p_1^{\text{Pack}} z^{132} + p_2^{\text{Pack}} z^{60} + p_3^{\text{Pack}} z^0 \quad (1.32)$$

$$L_z\{B6_{60}\} = p_1^{B6_{60}} z^{60} + p_2^{B6_{60}} z^0 \quad (1.33)$$

$$L_z\{B6_{21}\} = p_1^{B6_{21}} z^{21} + p_2^{B6_{21}} z^0 \quad (1.34)$$

When using CHB topology, cruise motors have nine branches of "BB"–"CHB" connected to them, whereas the small motors have six parallel branches each (refer to the RBD in Fig. 1.6).

$$L_z\{BB_{60}\} = p_1^{BB_{60}} z^{7.1} + p_2^{BB_{60}} z^{3.2} + p_3^{BB_{60}} z^0 \quad (1.35)$$

$$L_z\{BB_{21}\} = p_1^{BB_{21}} z^{3.8} + p_2^{BB_{21}} z^{1.7} + p_3^{BB_{21}} z^0 \quad (1.36)$$

$$L_z\{CHB_{60}\} = p_1^{CHB_{60}} z^{13.3} + p_2^{CHB_{60}} z^{6.67} + p_3^{CHB_{60}} z^0 \quad (1.37)$$

$$L_z\{CHB_{21}\} = p_1^{CHB_{21}} z^7 + p_2^{CHB_{21}} z^{3.5} + p_3^{CHB_{21}} z^0 \quad (1.38)$$

1.5.3 6P6 Drive Topology

The 6P6 topology is a more conventional design approach with two 60 kW six-phase cruise motors and four 31.5 kW six-phase motors, which can be seen in Fig. 1.10. The estimated total mass of this alternative is 78.8 kg. The L_z-transform of B6 inverters is presented in Eqs. (1.39)–(1.40), "BB" and "CHB" blocks are represented by expressions in (1.41)–(1.44), whereas the motors and propellers of 6P6 alternative are listed in Eqs. (1.45)–(1.48).

$$L_z\{B6_{60}\} = p_1^{B6_{60}} z^{60} + p_2^{B6_{60}} z^0 \quad (1.39)$$

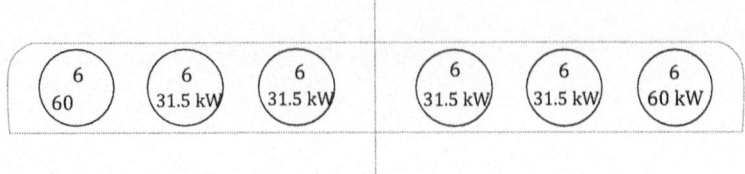

Fig. 1.10 6P6 representation consisting of 6 six-phase motors

$$L_z\{B6_{31}\} = p_1^{B6_{31}} z^{31.5} + p_2^{B6_{31}} z^0 \qquad (1.40)$$

$$L_z\{BB_{60}\} = p_1^{BB_{60}} z^{10.73} + p_2^{BB_{60}} z^{4.88} + p_3^{BB_{60}} z^0 \qquad (1.41)$$

$$L_z\{BB_{31}\} = p_1^{BB_{31}} z^{5.63} + p_2^{BB_{31}} z^{2.56} + p_3^{BB_{31}} z^0 \qquad (1.42)$$

$$L_z\{CHB_{60}\} = p_1^{CHB_{60}} z^{20} + p_2^{CHB_{60}} z^{10} + p_3^{CHB_{60}} z^0 \qquad (1.43)$$

$$L_z\{CHB_{31}\} = p_1^{CHB_{31}} z^{10.5} + p_2^{CHB_{31}} z^{5.25} + p_3^{CHB_{31}} z^0 \qquad (1.44)$$

$$L_z\{M6_{60}\} = p_1^{M6_{60}} z^{60} + p_2^{M6_{60}} z^{50} + p_3^{M6_{60}} z^{40} + p_4^{M6_{60}} z^{30} + p_5^{M6_{60}} z^0 \qquad (1.45)$$

$$L_z\{M6_{31}\} = p_1^{M6_{31}} z^{31.5} + p_2^{M6_{31}} z^{26.14} + p_3^{M6_{31}} z^{20.8} + p_4^{M6_{31}} z^{17.75} + p_5^{M6_{31}} z^0 \qquad (1.46)$$

$$L_z\{P_{60}\} = p_1^{P_{60}} z^{60} + p_2^{P_{60}} z^0 \qquad (1.47)$$

$$L_z\{P_{31}\} = p_1^{P_{31}} z^{31.5} + p_2^{P_{31}} z^0 \qquad (1.48)$$

1.5.4 9P4 Drive Topology

Lastly, with only four motors of 61.5 kW (see Fig. 1.11), this simple alternative involves only nine-phase motors, their total mass estimated at 78.8 kg. The L_z-transforms for elements of 9P4 are listed in Eqs. (1.49)–(1.53).

$$L_z\{B6_{9P4}\} = p_1^{B6_{9P4}} z^{61.5} + p_2^{B6_{9P4}} z^0 \qquad (1.49)$$

$$L_z\{BB_{9P4}\} = p_1^{BB_{9P4}} z^{7.3} + p_2^{BB_{9P4}} z^{3.3} + p_3^{BB_{9P4}} z^0 \qquad (1.50)$$

1.5 Proposed Alternative Topologies

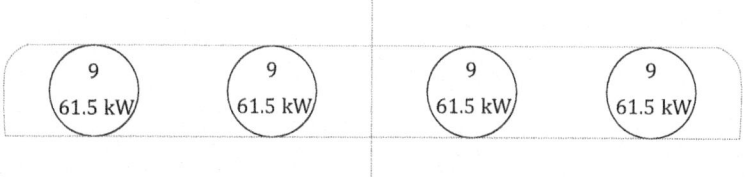

Fig. 1.11 9P4 representation, which consists of 4 nine-phase motors, two per wing side

$$L_z\{\text{CHB}_{9P4}\} = p_1^{\text{CHB}_{9P4}} z^{13.7} + p_2^{\text{CHB}_{9P4}} z^{6.8} + p_3^{\text{CHB}_{9P4}} z^0 \quad (1.51)$$

$$L_z\{\text{M9}_{9P4}\} = p_1^{\text{M9}_{9P4}} z^{61.5} + p_2^{\text{M9}_{9P4}} z^{54.1} + p_3^{\text{M9}_{9P4}} z^{47.3}$$
$$+ p_4^{\text{M9}_{9P4}} z^{40.6} + p_5^{\text{M9}_{9P4}} z^{33.8} + p_6^{\text{M9}_{9P4}} z^0 \quad (1.52)$$

$$L_z\{\text{P}_{9P4}\} = p_1^{\text{P}_{9P4}} z^{61.5} + p_2^{\text{P}_{9P4}} z^0 \quad (1.53)$$

1.5.5 Results

All alternative designs show improved availability for both demands and in the expected performance when compared to the X-57. It is noticeable that the alternatives with the CHB inverter topology provide higher drive train availability than the ones with B6-bridge in the Climb1 phase (see Fig. 1.12b). A similar trend is noticeable in the Roll phase with the exception of 6P6-B6 which has a higher availability than 9P4-CHB and 9P6PX-CHB alternatives (see Fig. 1.12a). When comparing the expected performance values, CHB alternatives had approximately 0.1 kW higher value than the B6 alternatives (Fig. 1.12c). In terms of availability, the 6P6-CHB configuration would be preferred over the other alternatives. When speaking of the total motor mass comparison, only the 9P6PX alternative has a lower estimated value (66.5 kg) than the 72.5 kg of the X-57 motors. Finally, adding the expected performance and the total motor mass into consideration, the 9P6PX-CHB could be considered as a slightly better design.

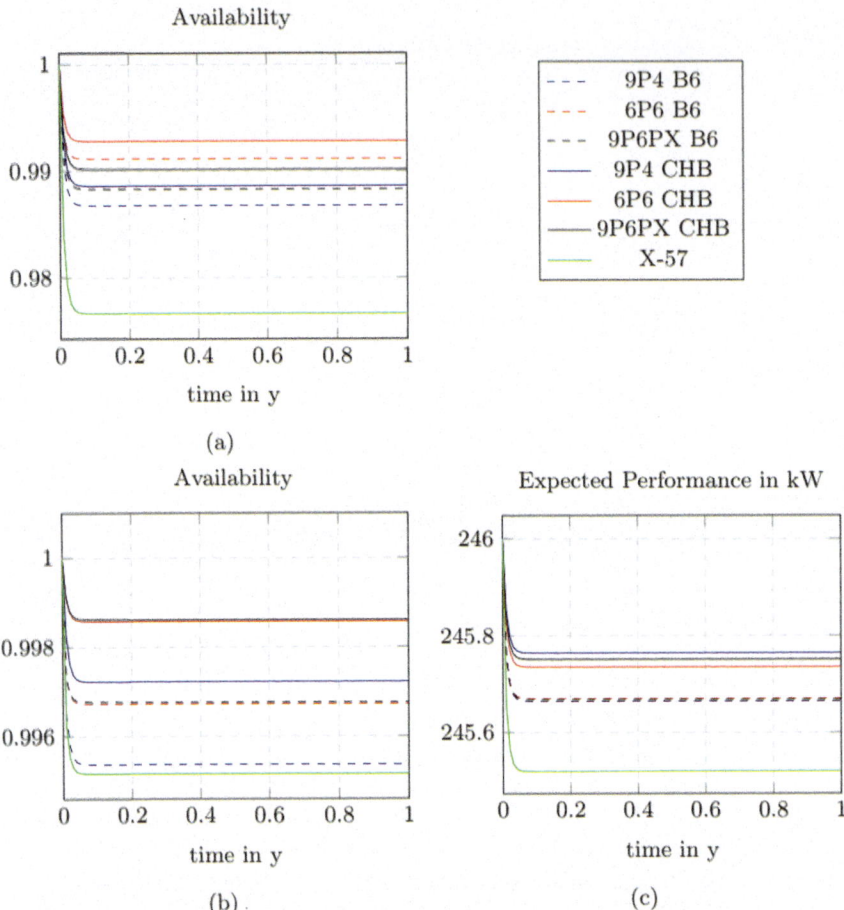

Fig. 1.12 Availability comparison at (**a**) Roll, $w = 240$ kW and (**b**) Climb1, $w = 210$ kW of different alternatives and X-57. Expected performance (kW) comparison of the alternatives (**c**)

1.6 Conclusion

The reliability analyses show that the replacement of X-57's three-phase motors with six-phase ones, with their suitable B6 inverters, as defined in the 6PX alternative, greatly improves availability of the electric drive train due to the use of fault-tolerant electric machines. Furthermore, all of the further proposed alternative designs (9P6PX, 6P6 and 9P4) have higher availability than the X-57 at the demands of 240 kW and 210 kW. Alternatives with the CHB inverter topology generally achieve higher availability values and higher expected performance than the B6 variants, due to the fault-tolerant inverter configuration. More specifically, 6P6 and

9P6PX with the CHB inverter topology stood out as the most reliable systems of this comparative analysis.

When analyzing the estimated mass of the alternatives, according to the stated specific power assumptions, the 9P6PX alternative was the only design with a lower total motor mass than one of the X-57, with a 6 kg advantage. However, if this mass advantage would be used to increase the battery mass, it would only bring about an insignificant increase in the flight range, with respect to the total battery mass of 390 kg. According to the presented statistical data, where the motors with a few dozen kW of power have a higher specific power than the ones closer to 100 kW, it is concluded that a distributed propulsion system with two larger multiphase cruise motors and a number of small multiphase motors used at roll and takeoff, provides a potential for a weight reduction, compared to conventional propulsion system designs.

There are some issues to be covered in future work with regard to this topic. First, the propulsion system can be analyzed with cruise motors having more than nine phases. Second, a more detailed Markov process description of the system components, including the aging process and scheduled maintenance as transition functions. Finally, a system analysis for non-repairable systems can be conducted for the drive train alternatives presented within this paper, since for the mentioned flight scenarios, a mid-operation repair is typically not feasible.

References

1. European Commission: 2030 Climate Target Plan, https://climate.ec.europa.eu/eu-action/european-green-deal/2030-climatetarget-plan_en. Accessed 03 Jul 2023
2. International Council on Clean Transportation: Performance analysis of regional electric aircraft. International Council on Clean Transportation (2022). https://theicct.org/publication/global-aviation-performanceanalysis-regional-electric-aircraft-jul22/. Accessed 10 May 2023
3. Future Flight: Eviation Aircraft Alice (2023), https://www.futureflight.aero/aircraft-program/alice. Accessed 03 Jul 2023
4. D. Hernandez-Lugo, S. Clarke, T. Miller et al., in *X-57 Maxwell Battery*, From cell level to system level design and testing (2018). https://ntrs.nasa.gov/api/citations/20180005737/downloads/20180005737.pdf. Accessed 03 Jul 2023
5. S. Clarke, SCEPTOR power system design: experimental electric propulsion system design and qualification for crewed flight testing, in *2016 16th AIAA Aviation Technology* (2016). https://ntrs.nasa.gov/api/citations/20160007822/downloads/20160007822.pdf. Accessed 03 Jul 2023
6. NASA, X-57 Maxwell (2021). https://www.nasa.gov/specials/X57/. Accessed 30 May 2023
7. B.T. Schiltgen, J. Freeman, ECO-150-300 design and performance: a tube-and-wing distributed electric propulsion airliner, in *AIAA Scitech 2019 Forum* (American Institute of Aeronautics and Astronautics, Reston, VA, 2019)
8. P. Schmollgruber, D. Donjat, M. Ridel et al, Multidisciplinary design and performance of the ONERA hybrid electric distributed propulsion concept (DRAGON), in *AIAA Scitech 2020 Forum* (American Institute of Aeronautics and Astronautics, Reston, VA, 2020)
9. V. Marciello, M. Di Stasio, M. Ruocco et al., Design exploration for sustainable regional hybrid-electric aircraft: a study based on technology forecasts. Aerospace **10**(2), 165 (2023). https://doi.org/10.3390/aerospace10020165

10. A. Lisnianski, L_z-transform for a discrete-state continuous-time Markov process and its applications to multi-state system reliability, in *Recent Advances in System Reliability*. ed. by A. Lisnianski, I. Frenkel (Springer, London, 2012), pp. 79–95
11. A. Lisnianski, Application of extended universal generating function technique to dynamic reliability analysis of a multi-state system. Paper presented at the second international symposium on stochastic models in reliability engineering, life science and operation management (SCE, Beer Sheva, 2016). https://doi.org/10.1109/SMRLO.2016.12
12. I. Frenkel, I. Bolvashenkov, L. Khvatskin et al., The Lz-transform method for the reliability and fault tolerance assessment of Norilsk-Type Ship's diesel-geared traction drives. Transp. Telecommun. **19**(4), 284–293 (2018). https://doi.org/10.2478/ttj-2018-0023
13. I. Frenkel, I. Bolvashenkov, H.-G. Herzog et al., L_z-transform approach for fault tolerance assessment of various traction drives topologies of hybrid-electric helicopter, in *Recent Advances in Multi-state Systems Reliability*. ed. by A. Lisnianski, I. Frenkel, A. Karagrigoriou (Springer, Cham, 2018), pp. 321–342
14. W. Zeng, G. Shen, B. Chen et al., Lz-transform method and markov reward approach for flexible manufacturing system performance evaluation. Appl. Sci. **9**(19), 4153 (2019)
15. A. Lisnianski, I. Frenkel, Y. Ding, *Multi-state System Reliability Analysis and Optimization for Engineers and Industrial Managers* (Springer, London, 2010)
16. G. Levitin, *The Universal Generating Function in Reliability Analysis and Optimization* (Springer, London, 2005)
17. NASA, Fact Sheet: NASA X-57 Maxwell, ed. M. Conner (2018). https://www.nasa.gov/centers/armstrong/news/FactSheets/FS109.html. Accessed 05 Apr 2023
18. J. Chin, *Battery-Performance-Modeling-on-Maxwell-X57* (2018). https://github.com/jcchin/Battery-Performance-Modelingon-Maxwell-X57/blob/master/missionTable.tex. Accessed 26 May 2023
19. A. Dubois, M. van der Geest, J. Bevirt et al., Design of an electric propulsion system for SCEPTOR's outboard nacelle. AIAA 2016-3925 (2016). https://doi.org/10.2514/6.2016-3925
20. D. Hall, C.J. Anderson et al., Development of a Maxwell X57 high lift motor reference design. AIAA (2019). https://doi.org/10.2514/6.2019-4481
21. S. Clarke, M. Redifer, K. Papathakis et al., X-57 power and command system design. Paper presented at 2017 IEEE Transportation Electrification Conference and Expo (ITEC) (IEEE, 2017), pp. 393–400
22. I. Bolvashenkov, J. Kammermann, H.-G. Herzog et al., Fault tolerance assessment of multi-motor electrical drives with multi-phase traction motors based on L_z-transform. Paper presented at 2019 fourteenth international conference on ecological vehicles and renewable energies (EVER) (IEEE, 2019), pp. 1–6
23. NASA, General aviation aircraft reliability study (2021). https://ntrs.nasa.gov/api/citations/20010027423/downloads/20010027423.pdf. Accessed 26 Mar 2023
24. I. Bolvashenkov, J. Kammermann, W. Zeng et al., Comparative reliability analysis of different traction drive topologies for a search-and rescue helicopter, in *Stochastic Models in Reliability Engineering*. ed. by L. Cui, I. Frenkel, A. Lisnianski (CRC Press, Boca Raton, 2020), pp. 331–354
25. H.D. Kim, A.T. Perry, P.J. Ansell, Review of distributed electric propulsion concepts for air vehicle technology. Paper presented at 2018 AIAA/IEEE electric aircraft technologies symposium (American Institute of Aeronautics and Astronautics, Reston, VA, 2018)
26. I. Bolvashenkov, J. Kammermann, H.-G. Herzog et al. Comparison of the battery energy storage and fuel cell energy source for the safety-critical drives considering reliability and fault tolerance, in *Proceedings of the International Conference on Information and Digital Technologies (IDT)* (IEEE, 2017), pp. 66–73
27. X. Shu, W. Yang, Y. Guo et al., A reliability study of electric vehicle battery from the perspective of power supply system. J. Power Sources **451**, 227805 (2020)
28. T. Donateo, L. Spada Chiodo, Design and reliability analysis of a series/parallel hybrid system with a rotary engine for safer ultralight aviation. Appl. Sci. **13**(7), 4155 (2023). https://doi.org/10.3390/app13074155

29. A. Perez, V. Quintero, H. Rozas et al., Modelling the degradation process of lithium-ion batteries when operating at erratic state-of-charge swing ranges. Paper presented at 2017 4th International Conference on Control, Decision and Information Technologies (CoDIT) (IEEE, 2017), pp. 860–865
30. S.-W. Eom, M.-K. Kim, I.-J. Kim et al., Life prediction and reliability assessment of lithium secondary batteries. J. Power Sources **174**(2), 954–958 (2007)
31. N. Williard, W. He, M. Osterman et al., Reliability and failure analysis of lithium ion batteries for electronic systems. Paper presented at 13th international conference on electronic packaging technology and high-density packaging (IEEE, 2012), pp. 1051–1055
32. P. Keil, *Aging of Lithium-Ion Batteries in Electric Vehicles*. Dissertation, Technical University Munich (2017). https://mediatum.ub.tum.de/doc/1355829/document.pdf. Accessed 28 May 2023
33. L. Cadwallader, Review of maintenance and repair times for components in technological facilities (2012), https://inldigitallibrary.inl.gov/sites/sti/sti/5554588.pdf. Accessed 23 Apr 2023
34. M.V. Bendarkar, D. Sarojini, E. Harrison et al., Evaluation of off nominal performance and reliability of a distributed electric propulsion aircraft during early design. AIAA 2021-1723 (2021). https://doi.org/10.2514/6.2021-1723
35. J. Colmenares, D.-P. Sadik, P. Hilber et al., Reliability analysis of a high-efficiency SiC three-phase inverter for motor drive applications. Paper presented at 2016 IEEE Applied Power Electronics Conference and Exposition (APEC) (IEEE, 2016), pp. 746–753
36. I. Bolvashenkov, J. Kammermann, T. Lahlou et al., Comparison and choice of a fault tolerant inverter topology for the traction drive of an electrical helicopter. Paper presented at 2016 International Conference on Electrical Systems for Aircraft, Railway, Ship Propulsion and Road Vehicles and International Transportation Electrification Conference (ESARS-ITEC) (2016), pp. 1–6
37. US Department of Defense, *MIL-HDBK-217F: Reliability Prediction of Electronic Equipment* (Department of Defense, Washington DC, 1991)
38. D. Hirschmann, D. Tissen, S. Schroder et al., Reliability prediction for inverters in hybrid electrical vehicles. IEEE Trans. Power Electron. **22**(6), 2511–2517 (2007)
39. R. Dobler, T. Schuhmann, R.B. Inderka et al., High performance drive for electric vehicles—system comparison between three and six phase permanent magnet synchronous machines. Paper presented at 18th European Conference on Power Electronics and Applications (EPE'16 ECCE Europe) (IEEE, 2016), pp. 1–10
40. M.A. Frikha, J. Croonen, K. Deepak et al., Multiphase motors and drive systems for electric vehicle powertrains: state of the art analysis and future trends. Energies **16**(2), 768 (2023)
41. G. El Murr, A. Griffo, J. Wang et al., Reliability assessment of fault tolerant permanent magnet AC drives. Paper presented at IECON 2015—41st Annual Conference of the IEEE Industrial Electronics Society (IEEE, 2015), pp. 2777–2782
42. T. Lahlou, Design and implementation of a 17-level cascaded H-bridge inverter for battery energy storage systems in the low voltage grid. PhD Thesis, Technical University Munich (2020), https://mediatum.ub.tum.de/doc/1455549/1455549.pdf. Accessed 23 May 2023
43. J. Kammermann, I. Bolvashenkov, J.L. Ugalde et al., Choice of phase number for a multi-phase traction motor to meet requirements on fault tolerance. Paper presented at International Conference on Electrotechnical Complexes and Systems (ICOECS) (2019), pp. 1–6
44. A. El-Refaie, M. Osama, High specific power electrical machines: a system perspective. Paper presented at 20th International Conference on Electrical Machines and Systems (ICEMS) (2017), pp. 1–6
45. S. Fang, H. Liu, H. Wang et al., High power density PMSM with lightweight structure and high-performance soft magnetic alloy core. IEEE Trans. Appl. Superconduct. **29**(2), 1–5 (2019)
46. S. Wu, J. Zhou, X. Zhang et al., Design and research on high power density motor of integrated motor drive system for electric vehicles. Energies **15**(10), 3542 (2022)
47. M. Meindl, X.J. Liu, F. Hilpert et al., Design proposal and optimization potential for an electric drive motor in a 50 PAX hybrid-electric regional aircraft application. Paper presented at 11th

International Conference on Power Electronics and ECCE Asia (ICPE 2023—ECCE Asia) (2023)
48. R.S. Da Rosa, L.A. Pereira, L.F.A. Pereira et al., Comparison of operating curves of five-phase and three-phase induction machines of same size. Paper presented at IECON 2014—40th Annual Conference of the IEEE Industrial Electronics Society (IEEE, 2014), pp. 450–455
49. A. Fleitas, M. Ayala, O. Gonzalez et al., Winding design and efficiency analysis of a nine-phase induction machine from a three-phase induction machine. Machines **10**(12), 1124 (2022)

Chapter 2
Fault-Tolerant Design and Control Strategies for Multiphase Traction Electric Drives

Abstract The chapter presents motor designs developed for the use in systems of multiphase induction motor drives with phase-pole control (which is a special case of over-phase control in accordance with the laws of energy efficiency acting in the field of multiphase induction motor drives. The authors of the paper developed three basic design options for these motors: a motor with a toroidal multiphase stator winding, a motor with a diamagnetic shield (screen) on the stator, and a motor with a multi-rod stator winding. The principles of construction of power supplies for motors with multi-rod stator winding have been developed. Further development of the concept of an induction motor with a multi-rod winding of the stator allowed the authors of the paper to develop principles for constructing windingless AC machines, which will have a simpler design and lower manufacturing cost than existing AC machines. A comparison is made of post-emergency control strategies that can be used in the field of multiphase phase-pole controlled induction motor drives.

Keywords Multiphase induction motor drive · Over-phase control · Fault-tolerant design · Windingless machine · Post-fault control strategies

2.1 Introduction

At present, it is a well-known fact that with an increase in the number of phases m of an induction motor drive to more than four, it becomes possible to improve a number of technical and economic characteristics of the drive system [1–14]. This is due, in particular, to the fact that when $m > 4$, it becomes possible to use some non-traditional methods of controlling an induction motor in the drive system, which, in principle, cannot be used when $m = 3$ and $m = 4$. These control methods include the over-phase control method (OPM). The essence of the OPM lies in the fact that when using it, the change in the rotational speed of the rotor and the torque on the rotor shaft is not achieved due to a change in the frequency or amplitude of the supply voltage. However, only the magnitude of the phase shift between the voltages of adjacent phases of the motor power source is increased by an integer factor, and therefore the motor (without any variation in the frequency and amplitude of this voltage).

OPM obtains two possibilities—over-synchronous control method (OSM) and phase-pole control one (PPM). The use of these methods allows to achieve the following effects:

- the increase of the motor rotor speed of rotation over its synchronous value (OSM) or
- the increase of the motor torque (PPM) without any change in the amplitude and frequency of the inverter output voltage (PPM) [5–8, 10].

To create PPM-controlled multiphase (i.e. with more than four phases) induction motor drives with high energy efficiency, in accordance with the laws of energy efficiency in force in the field of multiphase induction motor drives, it is required to develop special designs of induction motors and their corresponding power supplies. This article is devoted to the solution of this problem, as well as to the analysis of strategies for post-fault control of multiphase PPM-controlled induction motor drives.

2.2 The Features of Controlled Multiphase Electric Drives

2.2.1 Description of OPM

The main idea of the control according to OPM is that in this case the electrical angles α between the voltages (or currents) of the nearest (in time) phases of inverter are increased by a factor of H (in comparison with any traditional control method) without any change of the inverter voltage (or current) amplitude and frequency. This means in this case, $\alpha_H = H \cdot \alpha_T$, where H is some integer, α_T is the value α when some traditional control method is used ($\alpha_T = 2\pi/m$), and α_H is the value α when OPM is used [5–8, 10]. The range of the parameter H (including its maximal value), which can be achieved in the given drive system, depends on the phase number of the system and on the type of the stator winding.

The change of parameter H results in the change of the filtering properties of the generator multiphase induction motor drive. In particular, the numbers of the harmonics, which take part in the creation of the magnetic field in a motor air gap, are described by the following equation:

$$H \cdot c \pm n/p = b \cdot m, \qquad (2.1)$$

where c is the number of the phase voltage (or current) harmonics (i.e. the number of the time harmonics), n is the numbers of the harmonics of the functions which

2.2 The Features of Controlled Multiphase Electric Drives

describe a space distribution of the mutual inductances between motor phase windings (i.e. the numbers of the space harmonic), the coefficient $b = 0, \pm (1, 2, 3, \ldots)$, and p is the number of motor poles pairs.

There are two ways of OPM realization:

1. By the corresponding change in the switching algorithm of the inverter transistors without any changes in the traditional scheme of the inverter (i.e. without application of any electronic or mechanical switches of the motor stator phase windings or their sections). The first way can be used both for low- and high-power multiphase drive systems. This way can be used both for low- and high-power multiphase drive systems.
2. By the use of an electronic or mechanical phase commutator placed between a frequency converter and an induction motor [15]. In this case, the change in the OPM parameter H is not reached by any change in the electrical angles α between the voltages (or currents) of the nearest phases of inverter, but by the change in the version of the connection output terminals of the inverter to the terminals of the motor stator phase windings. Hence, the multiphase inverter has traditional scheme and operates at the sole value of the parameter $H = 1$. The change in the value of the parameter H is achieved by the abovementioned phase commutator placed between an inverter and a motor.

The second way is more expedient for the multi-motor drive systems, in which different motors fed by one frequency converter may operate at different values of parameter H at the same moment (for example, for distribution of traction forces between the motorized wheels of a wheel vehicle). The structural diagram of the drive system for this case is shown in Fig. 2.1, where FC is a frequency converter having an m-phase inverter of traditional type, PC is electronic or mechanical phase commutator, M is induction motor, and CS is the control system by the multi-motor drive. The electronic or mechanical phase commutator PC must connect the output terminals of inverter to the terminals of the motor stator phase windings according to the following algorithm:

$$i_M = N_1 - N_2, \tag{2.2}$$

where

$$N_1 = H \cdot (i_{in} - 1) + 1, \tag{2.3}$$

$$N_2 = \begin{cases} 0 & \text{if } H = 1, \\ m \cdot [(N_1 - 1)/m] & \text{if } H \geq 2, \end{cases} \tag{2.4}$$

where i_M is the number of a motor stator phase winding, i_{in} is the number of an inverter phase, and $[(N_1 - 1)/m]$ is the integer part of the number $(N_1 - 1)/m$.

Fig. 2.1 Structural diagram of the multi-motor OPM-controlled drive system with electronic or mechanical phase commutators between frequency converter and induction motors

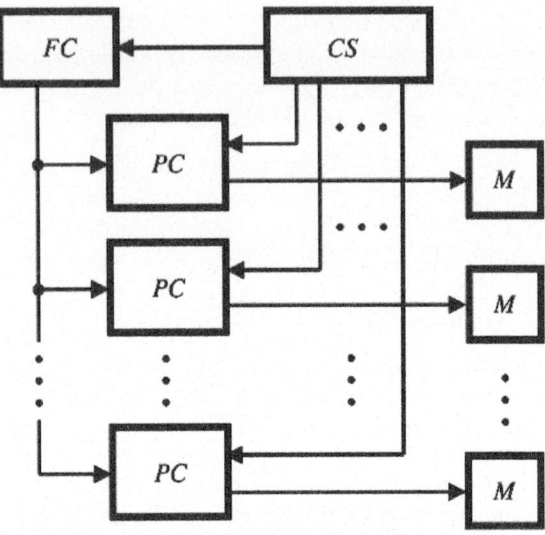

The peculiarities of the phase commutator PC may be conveniently shown by the example of the following case (see Fig. 2.2, where S_h is the source of the signal h, and TC are the electronic or mechanical terminal changers):

- the drive system is single-motor, $m = 6$,

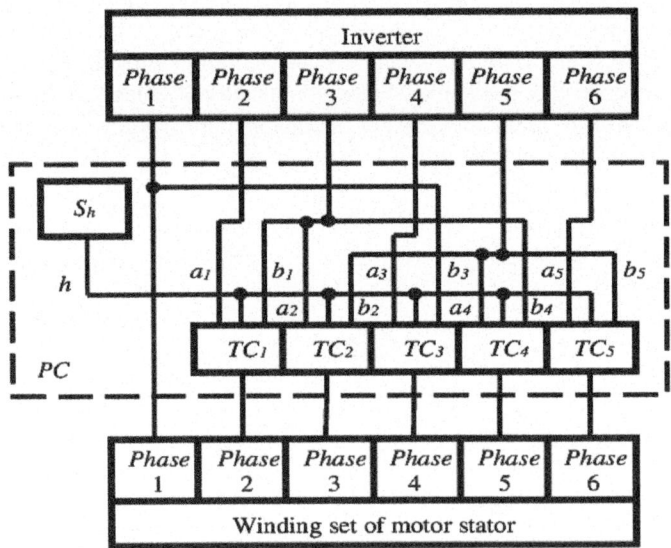

Fig. 2.2 Structural diagram of the 6-phase single-motor OPM-controlled system "inverter–induction motor" with phase commutator between inverter and induction motor

2.2 The Features of Controlled Multiphase Electric Drives

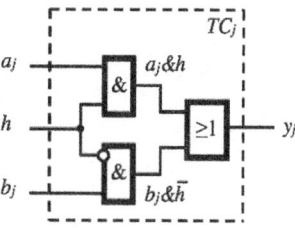

Fig. 2.3 Logic version of the electronic terminal changer, where $j \in [1; 5]$

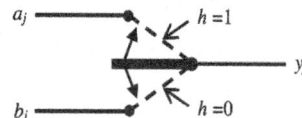

Fig. 2.4 Mechanical switching analogue of the logic scheme shown in Fig. 2.3, where $j \in [1; 5]$

- 6-phase inverter has traditional scheme and operates at the sole value of the parameter H: $H = 1$,
- the change in the parameter H value is achieved by phase commutator PC,
- the parameter H may be equal to 1 or 2,
- the logic signal h depends on the value of the OPM parameter H (i.e. $h = 1$ if $H = 1$, and $h = 0$ if $H = 2$),
- and TC_j is the electronic or mechanical terminal changer for the phase having number $j = i + 1$, $i \in [1; m]$, $j \in [1; m-1]$.

If the drive system is low-powered, the type of electronic terminal changers (TC) may be logic. The electronic logic version of the terminal changer TC_j is shown in Fig. 2.3 (for the case of low-power drive system), where the unit "&" is an AND-type logic gate, and the unit "≥ 1" is an OR-type logic gate. The mechanical switching analogue of the logic scheme is shown in Fig. 2.3 and presented in Fig. 2.4.

These schemes carry out the following logic function:

$$y_j = a_j \,\&\, h \vee b_j \,\&\, \bar{h} = \begin{cases} a_j & \text{if } h = 1; \\ b_j & \text{if } h = 0. \end{cases} \quad (2.5)$$

The coefficient H is the major parameter of OPM that characterizes the type of this control method (the value $H = 1$ corresponds to any traditional control method, and the value $H > 1$ corresponds to OPM).

OSM differs from PPM in the following: when OSM is used, the spectrum of the function, which describes the space distribution of the mutual inductances between motor phase windings, does not change for all values of H, which have to be realized in the given drive system. When PPM is used, the spectrum of the abovementioned function changes during H-value changing process.

The range of the parameter H (including its maximal value), which can be achieved in the given drive system, depends on the phase number of the system and on a motor stator winding type.

Application of OSM and PPM was described in [5–8, 10].

2.2.2 Possibility to Decrease the Step of Speed Changing

The main disadvantage of the proposed drive system is the stepwise change of the motor speed. However, it is possible to obtain some quasi-continuous change of this speed by the decrease in the step of its change. This speed change step decrease can be obtained by the use of the following methods:

Method 1. By the choice of some values of the parameters $H_{MIN} \gg 1$ and $H_{NOM} \gg 1$ (it is always $H_{NOM} > H_{MIN}$) as the minimum and nominal values of the parameter H. Value H_{MIN} corresponds to the maximum speed of rotation of the motor rotor, and the value of H_{NOM} corresponds to the nominal speed of rotation of the motor rotor, which was chosen by the designer of the motor and the drive system as a whole.

For example, the abovementioned speed change step decrease may be characterized by coefficient w_S/w_{S+1}, where w_S is the rotor speed of rotation before the change of the parameter H, w_{S+1} is this speed after the change of the parameter H by one step (by one unit), and S is the step of the parameter H change. The values of the coefficient w_S / w_{S+1} are shown in Table 2.1 for the case when $m = 60$ and the parameter H changes from its least value (i.e. $H_{MIN} = 1$) to $H = 5$, where H_{NOM} is the value of the parameter H chosen as the nominal value of this parameter.

An increase in $H_{NOM} > 1$ inevitably leads to the need to increase the frequency of the supply voltage by a factor of H_{NOM}.

Method 2. By the use of not 1-, but 2-stage (2-level) commutator [15] shown in Fig. 2.5, where Inv. is a multiphase inverter, M is multiphase induction motor, CU is control unit, S-1 is the 1st stage of PC, and S-2 is the 2nd stage of PC. In this case, the 1st commutator stage implements the transition to PPM, and the 2nd commutator stage implements the transition to OSM. The combined operation of both commutator stages will allow to decrease the motor speed change step.

Method 3. By the combined use of the abovementioned two methods (Method 1 and Method 2).

The following should be noted here. The increase of the phase number of asynchronous electric drive system allows not only to improve a number of its technical and economic characteristics, but also to create the hybrid traction drives for different

Table 2.1 Values of the coefficient w_S/w_{S+1} for the case when $m = 60$ and the parameter H changes from its least value (i.e. $H = 1$) to $H = 5$

S	H_{NOM}				
	1	2	3	4	5
1	2.0	1.5	1.(3)	1.25	1.2
2	1.5	1.(3)	1.25	1.2	–
3	1.(3)	1.25	1.2	–	–
4	1.25	1.2	–	–	–
5	1.2	–	–	–	–

2.3 Design Implementation of the Law of *H*-Invariance

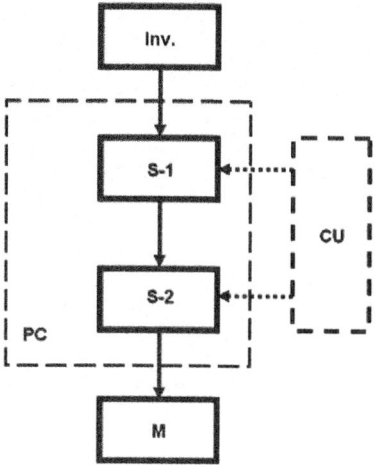

Fig. 2.5 Functional diagram of the power section of the proposed drive system having 2-stage commutator

vehicles according to scheme "Diesel-generator–induction motor" without application of any power electronic elements operating with high frequency, which will differ fundamentally from the existing ones, and have more simple design and control unit and less energy losses (in comparison with analogous existing hybrid drives).

2.3 Design Implementation of the Law of *H*-Invariance

2.3.1 On the Need to Fulfill the Law of H-Invariance

The law of *H*-invariance, described in details in [16, 17] reduces to the following two conditions, which must be fulfilled to achieve the PPM application opportunity:

1. The space harmonics having numbers $n = p \cdot H$ must be contained in the spectrum of the function describing the space distribution of the mutual inductances between motor phase windings for all parameter H values which have to be realized in the given drive system.
2. When $H > 1$, the amplitudes of the abovementioned space harmonics (i.e. having numbers $n = p \cdot H$) must have the values being not less than those at $H = 1$.

The abovementioned rules of efficiency invariance are particular cases of the fundamental principle, which prevails in the field of multiphase electric drives and may be stated as follows: to ensure a maximum energy efficiency of a multiphase electric drive system, the laws of space-temporal spectral relations, which act in the field of these systems, two steps should be fulfilled during the process of development. These two steps are the structural elements design and motor control mode elaboration.

Fig. 2.6 Mechanical characteristics of some multiphase phase-pole controlled induction motor with $m \geq 6$ for the cases when $H = 1$ (line 1) and $H = 2$ (lines 2 and 3)

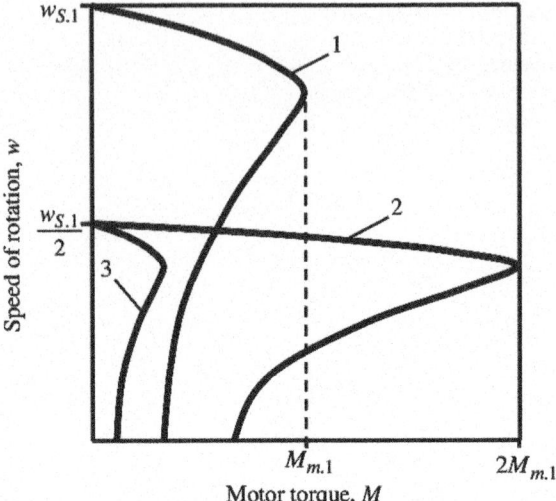

It is obvious that the fulfillment of these conditions (i.e. the abovementioned law of H-invariance) may be provided only by obtaining the galvanic decoupling between the induction motor stator slots, which cannot be achieved if the design of the motor and its stator winding set are traditional.

Because of this, it is necessary to use special designs of multiphase induction motors to ensure fulfillment of the H-invariance law [5–8, 10, 18]. In support of the H-invariance law the mechanical characteristics (i.e. w–M characteristics) of some multiphase phase-pole controlled induction motor with $m \geq 6$ are presented in Fig. 2.6 for the cases when $H = 1$ (see line 1) and $H = 2$ (see lines 2 and 3), where line 2 is the mechanical characteristic if the law of H-invariance is fulfilled, line 3 the mechanical characteristic if the law of H-invariance is not fulfilled, w is the rotational speed, and M is the motor torque.

According to the space-temporal spectral relations and energy efficiency invariance laws acting in the inverter-fed multiphase induction motor drives [5–8, 10, 18], the following three design versions were developed by the authors of this chapter:

Design 1 (D1). The stator of the induction machine has a toroidal-type winding set. The rotor envelops the stator both on the outside and on the inside (when the motor is of rotating type) or from above and from below (when the motor is linear or bow-shaped).

Design 2 (D2). The induction motor stator core (i.e. magnetic circuit) consists of two parts magnetically separated from each other by diamagnetic shield.

Design 3 (D3). The stator winding set of the induction motor consists of the individual rods located in stator slots and insulated from stator magnetic circuit.

The application of any one of these design versions allows to achieve the fulfillment of the H-invariance law.

2.3 Design Implementation of the Law of *H*-Invariance

Below, the abovementioned three basic design versions (D1, D2, and D3) are illustrated by concrete examples.

2.3.2 Design 1 (D1)

The design of multiphase induction motor of rotating type worked out according to the basic design version D1 is shown in section in Fig. 2.7. This motor has toroidal-type stator winding set and E-shaped rotor enveloping the motor stator both on the outside and on the inside. The motor stator has two rows of slots: the first row is located on the inner surface of the stator core (the inner stator slots), and the second row is located on the outer surface of it (the outer stator slots). The slots disposition on the stator surface of the 8-phase induction motor developed according to the basic design version D1 is shown in Fig. 2.8. Every stator phase winding drops in radial direction in two slots—in one inner slot and in one outer slot.

Exterior view of the 24-phase induction motor stator, which is developing according to the drawing shown in Fig. 2.7, is shown in Fig. 2.9.

The various versions of the E-shaped rotor design of such induction motor can be used. For example, this rotor can be laminated (Fig. 2.10) or may have solid ferromagnetic body with circular or U-shaped axial-and-radial slots (Fig. 2.11) made on the rotor surface for the decrease of the electrical losses in the rotor circuit.

Analogous design can be used in the case where motor stator winding set is of any usual type (for example, winding set of drum type). In this case, such an

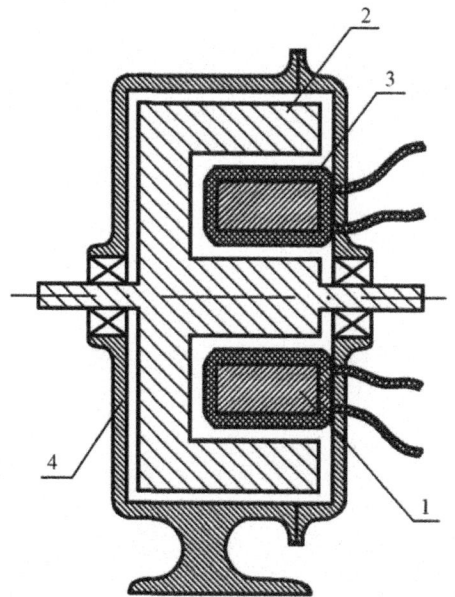

Fig. 2.7 Section of multiphase induction motor of rotating type worked out according to the basic design version D1, where 1 is the stator core (i.e. magnetic circuit), 2 is the rotor core (magnetic circuit), 3 is a stator phase winding, and 4 is the motor case

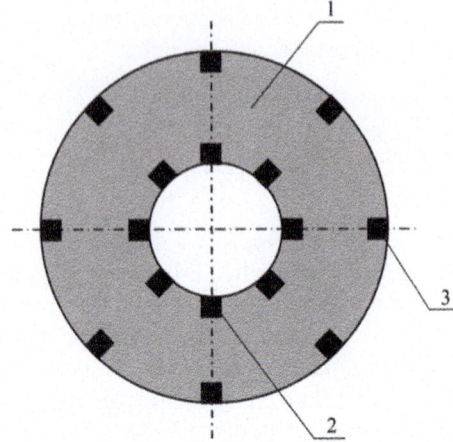

Fig. 2.8 Slots disposition on the stator surface of the 8-phase induction motor developed according to the basic design version D1, where 1 is the stator core (i.e. stator magnetic circuit), 2 is the inner stator slot, and 3 is an outer stator slot

Fig. 2.9 Exterior view of the 24-phase induction motor stator developing according to the drawing shown in Fig. 2.7

induction motor design version will allow to decrease the magnetic losses in the motor (Fig. 2.12).

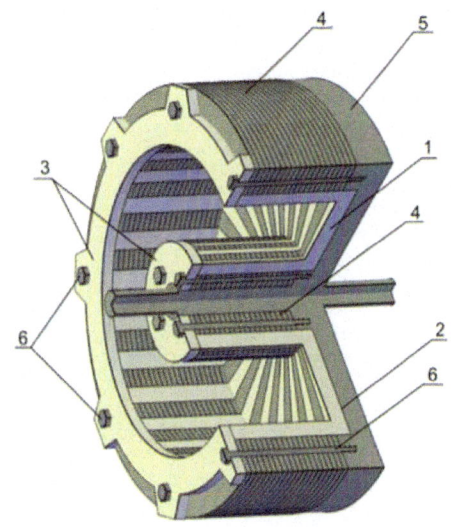

Fig. 2.10 E-shaped laminated rotor, where 1 and 2 are U-shaped axial-and-radial rods, 3 are short-circuited rings, 4 is the laminated part of rotor base, 5 is the solid part of rotor base, 6 are studs

2.3.3 Design 2 (D2)

The stator of a 6-phase flat type AC linear motor, which is worked out according to the basic design version D2 is presented in Fig. 2.13. In this case, every stator phase winding drops in two slots separated from each other by diamagnetic shield *DS*. The diamagnetic shield DS divides the magnetic field created by the stator winding into two subfields separated in space. The change of magnetic field pole number on space intervals between the borders of every abovementioned subfield is observed during PPM application process. The rotational version of this motor is shown in Fig. 2.14.

2.4 Design of the Drive System with Induction Motor

2.4.1 Induction Motor of Design Version D3

According to the basic design version D3, the induction motor stator winding set consists of the individual rods located in stator slots which are insulated from the stator magnetic circuit [5–8, 10, 18]. The rods can be made from copper, aluminum, etc. Every rod is connected to an individual phase of a frequency converter by one of its ends (Figs. 2.15, 2.16 and 2.17). Other rods ends are connected in star having or not having a neutral conductor. The rods of the stator winding are connected to the frequency converter via an m-phase transformer stepping down voltage. In this case, m is the number of the rods (i.e. the number of real phases of the drive system). Such winding design is simpler than the traditional ones, but in this case, it is necessary to

Fig. 2.11 E-shaped solid rotor, where **a** is E-shaped solid rotor having U-shaped axial-and-radial slots, **b** is E-shaped solid rotor having circular slots, 1 is the rotor body, 2 are U-shaped axial-and-radial rods, and 3 are circular slots

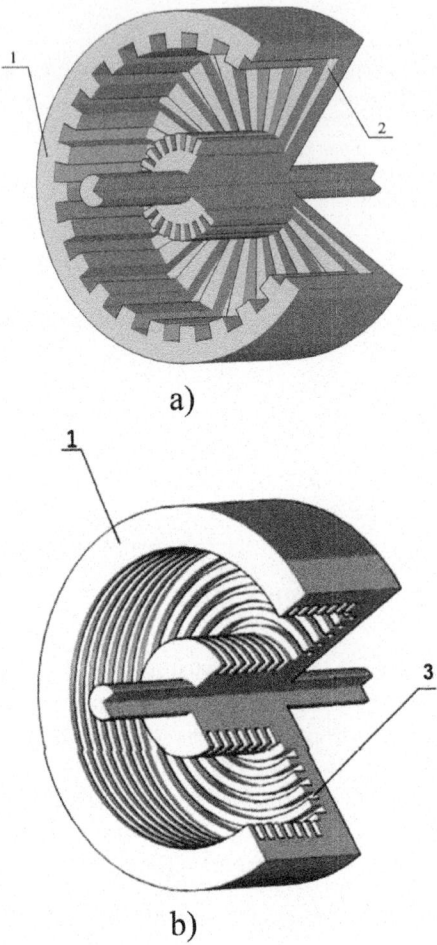

use transformer stepping down voltage, which has large mass and overall dimensions. This design version may be used for both linear and nonlinear induction motors.

2.4.2 Alternative Version of Connection of D3 Design Motor to Energy Source

Traditionally, multiphase inverters are built in such a way that the individual phases of the inverter are connected in parallel with each other (hereinafter—option *A*, Fig. 2.18). In this case, with an increase in the number of phases, the phase currents of the inverter decrease (at a constant phase voltage), which is one of the main advantages of increasing the number of phases of the inverter system to more than

2.4 Design of the Drive System with Induction Motor

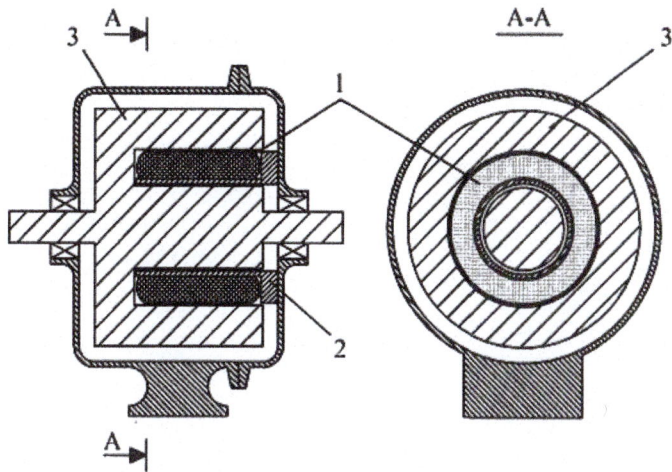

Fig. 2.12 Induction motor having the E-shaped rotor and a stator winding set of any usual type, but not having magnetic circuit of stator, where 1 is the stator winding set of usual (drum) type, 2 is the thin and durable plastic frame (with protrusions for fixing (laying) the winding) on which the stator winding is attached, and 3 is the E-shaped rotor core (magnetic circuit of whole motor)

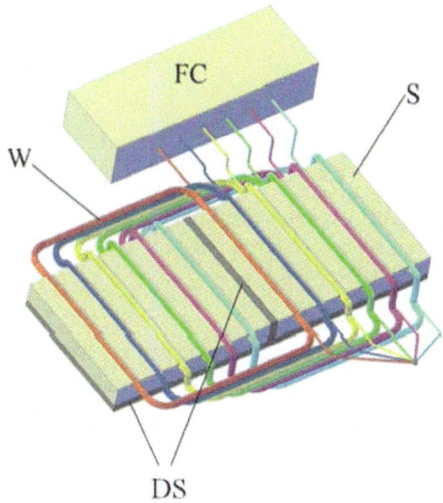

Fig. 2.13 Design version of the stator of a 6-phase flat type AC linear motor and winding connections, where FC is a 6-phase frequency converter, S is the motor stator, W is phase multiple-turn winding of the motor stator, and DS is diamagnetic shield

three, since this makes it possible to build inverters based on semiconductor devices (power transistors, thyristors, and diodes), designed for lower nominal current values than with the number of phases equal to three, and at the same time avoid their parallel connection, which in most cases leads to an increase in manufacturing cost and mass and overall dimensions of the inverter.

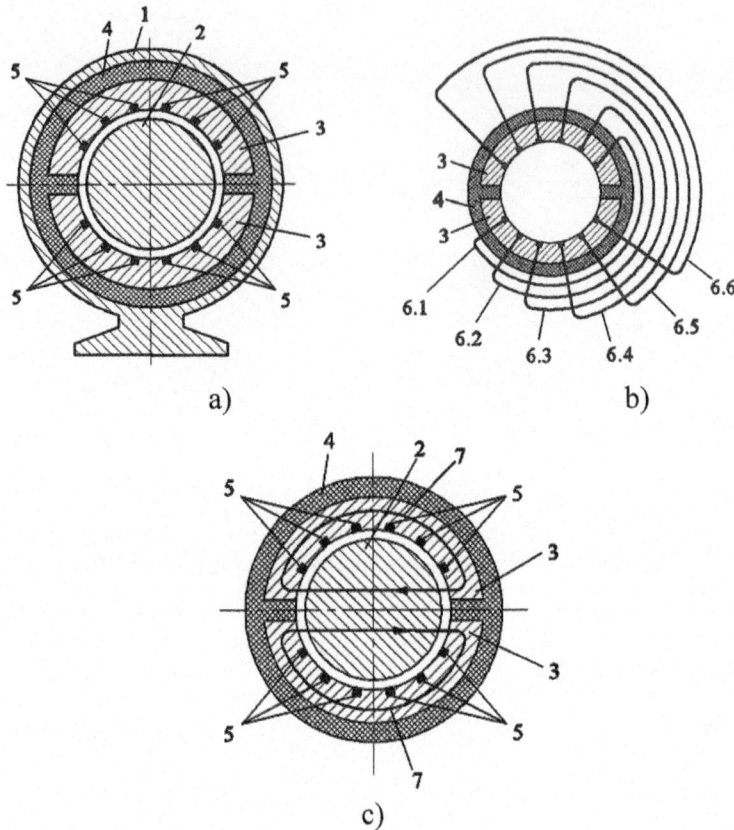

Fig. 2.14 Rotating version of D2, where 1 is the motor housing, 2 is the motor rotor, 3 is the stator core, 4 is the diamagnetic shield, 5 are the stator slots, 6.1–6.6 are the stator phase windings, and 7 is magnetic field in the motor

However, for the developed D3 design induction motor drive system, the opposite characteristic appears: it will have large phase currents at low values of phase voltages due to low values of active, reactive (inductive), and, consequently, the total electrical resistance of the phase windings of the motor stator (due to the specifics of the design of the rod winding of the motor).

In this case, it is advisable to apply the following variant of constructing a multi-phase inverter, alternative (in a certain sense) to the traditional variant described above. With the proposed option, a multiphase inverter consists of single-phase inverters connected in series (hereinafter—option B, Fig. 2.19 at $m' = 1$). In this case, the individual phases of the multi-phase inverter are also connected in series in the overall inverter circuit. Hence, an effect opposite to that characteristic of option A is observed: with this principle of constructing an inverter, with an increase in the number of phases, the phase currents of the inverter do not change, and the phase

2.4 Design of the Drive System with Induction Motor

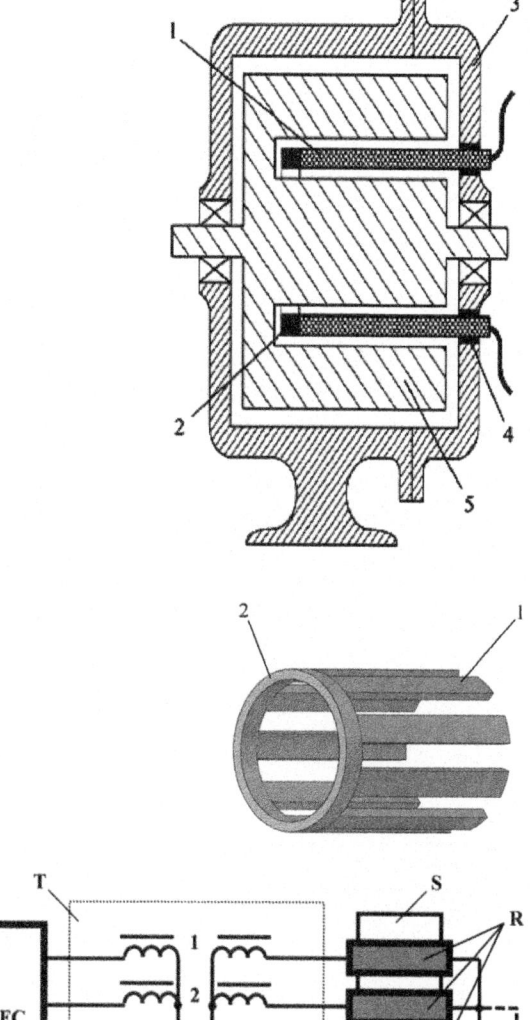

Fig. 2.15 Section of multiphase induction motor of rotating type worked out according to the basic design version D3, where 1 are the rods of the stator winding sets (the stator has not any magnetic circuit), 2 is the short-circuited ring, 3 is the motor housing, 4 are insulators, and 5 is the E-shaped rotor core (magnetic circuit)

Fig. 2.16 Stator of multiphase induction motor of rotating type worked out according to the basic design version D3, where 1 are the rods of the stator winding sets (the stator has not any magnetic circuit), and 2 is the short-circuited ring

Fig. 2.17 Connection of m-rod stator winding to a m-phase frequency converter according to D3, where FC is the frequency converter, T is an m-phase transformer stepping down voltage, S is the stator of an induction motor, R are the rods of the stator winding, and NC are neutral conductors

Fig. 2.18 Typical simplified circuit diagram of a 3-phase inverter

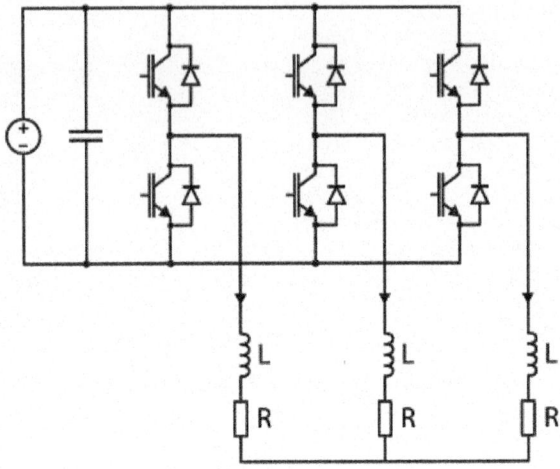

voltages decrease in proportion to the number of phases. In some cases, this feature may turn out to be an advantage of such a principle for constructing multiphase inverters, since it allows to build inverters with more than three phases based on semiconductor devices designed for lower voltages than in the case of a 3-phase version.

A third combined version of building multiphase inverters is also possible, in which the two principles described above are combined (option C): we are talking

Fig. 2.19 Functional diagram of inverter, corresponding to options B and C, where R is the rectifier, RF is the ripple filter, ISS is the m'-phase inverter subsystem, SSPW is the m'-phase subsystem of phase windings of the motor stator, Inv. is the inverter, and SW is the m-phase system of phase windings of the stator ($m = N \cdot m'$)

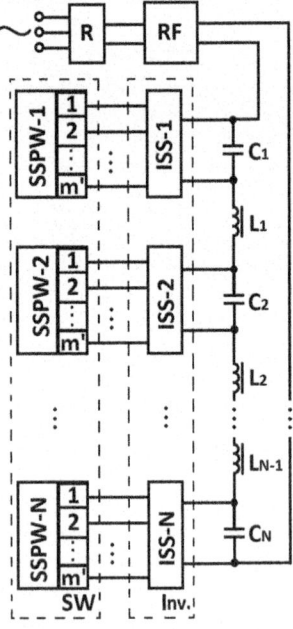

about the series–parallel principle of building multiphase inverters, in which an m-phase inverter is a series connection of N inverters, each having a number of phases equal to m', where $m' = m/N$. Limiting chokes L_i (where $i \in [1; (N-1)]$) are installed between adjacent m'-phase inverters (in series with them) to reduce the reactive effect of these inverters on each other (i.e. mutual influence on the variable component of the input current of each of these inverters). At the input of each of the m'-phase inverters, a capacitor C_j (where $i \in [1; N]$) is installed to provide a closed circuit for the flow of the reactive (alternating) component of the input current of this inverter (Fig. 2.19 at $m' > 1$). In such a circuit, the current I consumed by the entire m-phase circuit at even values of m is determined by the formula

$$I = U \cdot m'/(4 \cdot N \cdot R_f), \qquad (2.6)$$

where U is the voltage at the input of the m-phase inverter and R_f is the active resistance of the load phase of the inverter.

For odd values m formula (2.6) takes the following form:

$$I = U[(m')^2 - 1]/(4 \cdot N \cdot R_f \cdot m'). \qquad (2.7)$$

In this case, the phase voltage U_f of m-phase inverter (for both even and odd values of m') is determined by the formula

$$U_f = U/(2N). \qquad (2.8)$$

In general case, in such a system m'-phase inverters can have different number of phases. If necessary, step-down transformers can be connected between the ISS and the SSPW, as shown in Fig. 2.17.

It should be noted that the option B can be considered as a special case of option C, which seems to be the most promising due to its flexibility, which lies in the fact that with this principle of constructing an inverter, the designer has maximum scope for varying the values of phase currents and voltages, as well as for minimizing the cost of the designed system, based on from the real possibilities of the element base at his disposal.

2.5 Further Development of AC Machine Concept with Stator Rod Winding

2.5.1 Tubular-Rod Winding Filled with a Conductive Gaseous or Liquid Substance

In the first half of the twentieth century, Nikola Tesla conducted experiments with rarefied gases and found that these gases have a high electrical conductivity [19–25].

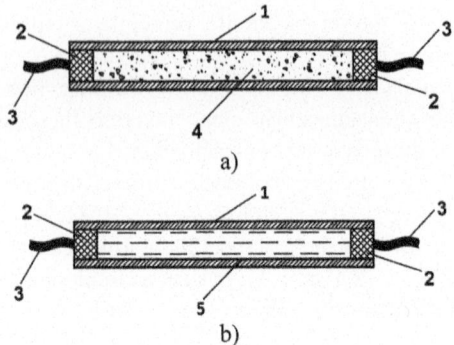

Fig. 2.20 Rod tubes filled with rarefied gas (**a**) and electrolyte (**b**), where 1 is a tube made of a material with high electrical insulating properties, 2 is an electrode, 3 is a wire (conductor), 4 is a rarefied gas, and 5 is an electrolyte

In this regard, it seems appropriate instead of the metal rods of the winding of the D3 design motor to use tubes made of a material with high electrical insulating properties and filled with some kind of rarefied gas, for example, rarefied air (Fig. 2.20a). In this case, the design of the entire multiphase winding of the stator of the D3 design motor can have the form shown in Figs. 2.21 and 2.22.

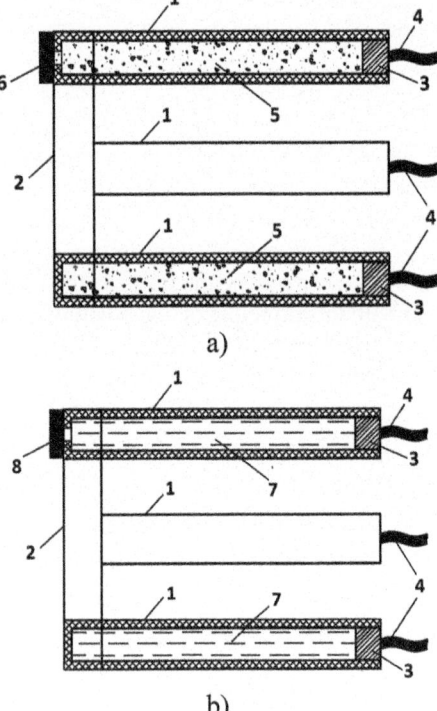

Fig. 2.21 Tubular-rod multiphase stator winding of the D3 design motor, filled with rarefied gas (**a**) and electrolyte (**b**), where 1 is a tube-rod, 2 is a short-circuiting ring-tube, 3 is an electrode, 4 is a wire (conductor), 5 is a rarefied gas, 6 is discharge valve, 7 is an electrolyte, and 8 is inlet valve

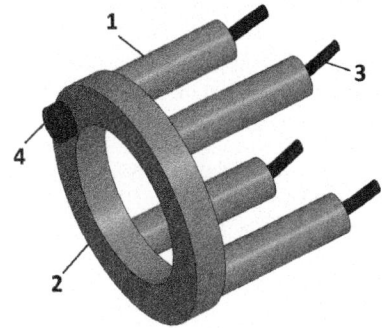

Fig. 2.22 3D model of the tubular-rod multiphase stator winding of the D3 design motor, filled with a rarefied gas or electrolyte, where 1 is a tube-rod, 2 is a short-circuiting ring-tube, 3 is a wire (conductor), and 4 is a discharge valve (for Fig. 2.21a) or inlet valve (for Fig. 2.21b)

Mass production and widespread use in practice of the D3 design motors with tubular-rod stator windings will eventually lead to the following positive consequences:

- Saving aluminum and expensive copper;
- Improvement of the environmental situation in the world due to the reduction of the volumes of environmentally harmful production of copper and aluminum required by industry and transport.

The electrical conductivity of rarefied gases depends on the frequency of the applied voltage: with an increase in this frequency, the conductivity of a rarefied gas increases [23–25]. This phenomenon causes the fact that in the field of phase-pole controlled AC motors, in which rarefied gases are used as conductors of electric current, it is most expedient to use methods 1 and 3 of reducing the step of discrete change in the rotor speed, described in Sect. 2.2.

In addition to rarefied gases, various electrolytes can be used in the tubular-rods of such a stator winding (Figs. 2.20b and 2.21b). Electrolytes are liquid conductors. These include, in particular, sea, lake, and river water (but not distilled water).

2.5.2 Windingless AC Motors and Generators

If the AC machine is designed to operate in a conductive medium, then it seems logical and expedient to use this medium as a substitute for individual structural elements of the stator and (or) rotor windings of the AC machine. Such conductive media can be the upper layers of the atmosphere (where there is highly rarefied air with high electrical conductivity [20, 21]) or an aqueous medium (sea, lake or river, but not distilled water).

For such a situation, the design of an asynchronous machine is shown in Fig. 2.23.

In such a machine, all channels and all the space around the stator and rotor will be filled with a gaseous or liquid conductive medium. When contacts 7 are connected to a source of multiphase alternating voltage (for example, an inverter),

Fig. 2.23 Section of a windingless AC machine, where 1 is the machine body, 2 is the side cover of the case, 3 is the laminated stator, 4 is the laminated rotor, 5 is the rotor shaft, 6 is the bearing, 7 is the wire (conductor), 8 is the through hole, and 9 are channels

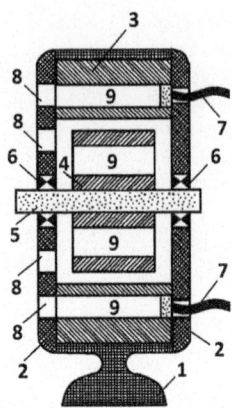

the conductive medium filling channels 9 will play the role of the rods of the stator and rotor windings of the D3 design motor. The conductive medium surrounding the stator and rotor of the machine will play the role of short-circuiting rings. The channels 9 can be covered from the inside with a layer of electrically insulating material. The machine can operate in both motor and generator mode.

Figure 2.24 shows an explosion diagram of the windingless AC machine shown in Fig. 2.23. In such a windingless AC machine, rotors of various designs can be used. Three examples are presented in Fig. 2.25. When using a smooth massive rotor (Fig. 2.25a), the windingless AC machine will be an asynchronous machine. When using the rotors shown in Fig. 2.25b and c, the machine will be a synchronous machine (additionally in the case, when using the rotor shown in Fig. 2.25b, the machine will be able to work only in the motor mode, i.e. will be a synchronous reluctance motor).

These types of AC machines have a wide field of application, for instance wind turbines designed to operate in the upper atmosphere, hydro generators, as well as electric generators to convert energy of sea waves and currents or tides.

Fig. 2.24 Explosion diagram of a windingless AC machine shown in Fig. 2.23, where 1 is the machine body with the stator, 2 is the rotor, and 3 is the body cover

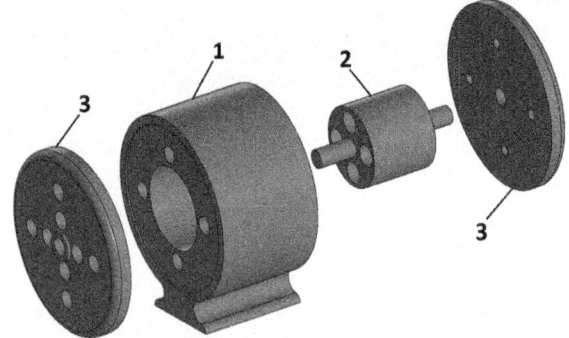

2.5 Further Development of AC Machine Concept with Stator Rod Winding

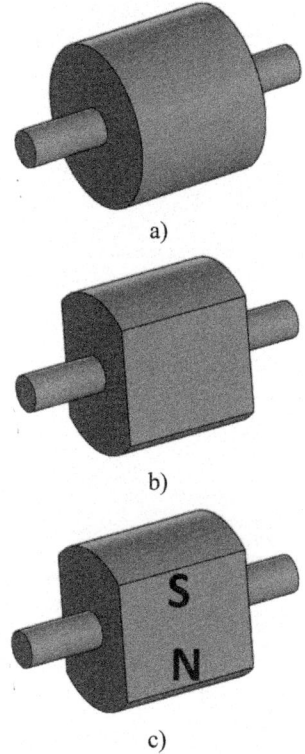

Fig. 2.25 Possible options for the rotor of a windingless AC machine shown in Figs. 2.23 and 2.24, where **a** is a solid smooth rotor, **b** is a laminated or solid rotor made of ferromagnetic material not having residual magnetization, **c** is a rotor that is a permanent magnet

The AC machine presented above can also be used on land below the upper layers of the atmosphere (both as a motor and as a generator), but for this it must be placed in a hermetically sealed chamber filled with either a rarefied gas (for example, rarefied air) or an electrolyte (Fig. 2.26).

Obviously, such an AC machine will be much cheaper than similar machines of the traditional type (especially if rarefied air will be used as a conductive medium surrounding this machine in a hermetically sealed chamber).

The AC machines presented above can also be used as induction motors in the PPM-controlled induction motor drive systems.

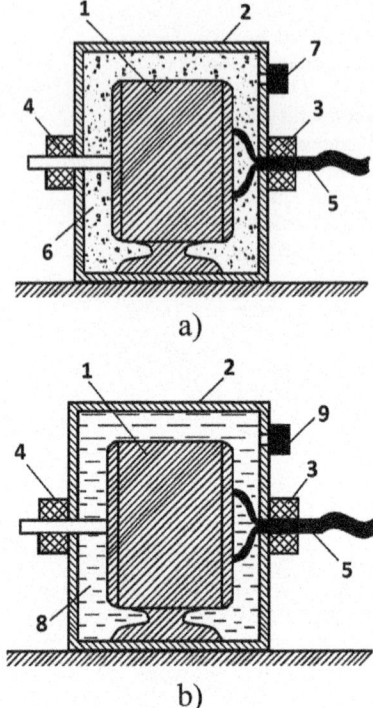

Fig. 2.26 Windingless AC machine placed in a hermetically sealed chamber filled with rarefied gas (**a**) and electrolyte (**b**), where 1 is a windingless AC machine, 2 is a hermetically sealed chamber, 3 is a seal, 4 is a shafting seal, 5 is an electric cable, 6 is a rarefied gas, 7 is a discharge valve, 8 is electrolyte, and 9 is an inlet valve

2.6 Additional Advantages of Multiphase Induction Motor Drives

2.6.1 Strategies to Increase Reliability of Multiphase Induction Motor Drives

The increase of the phase number of induction motor drive system allows to improve a number of technical and economic characteristics of this system [1–14], including to increase its reliability [5, 11–14, 26, 27].

Taken alone, the increase of the phase number to more than four (without any changes in the system of inverter phase voltages) does not ensure any essential increase of the electric drive reliability [26]. The increase of the phase number more than four allows to increase the reliability of electric drives only in the two following cases, when the following post-fault control strategies (PFCS) are used:

PFCS1. If in post-fault situations the output power of a multiphase (i.e. having $m > 4$) inverter per phase can be increased by the factor N_P (without the application of PPM in post-fault situations) by the corresponding increase in phase voltages and currents, where $N_P \approx m/m_N = m/(m - m_F)$, m_N is the number of the intact phases of an inverter, $m_N = m - m_F$, m_F is the number of the damaged phases of an inverter [28].

The application of this strategy is accompanied by the increase of mass and overall dimensions and manufacturing cost of the inverter-fed induction motor systems. This increase degree is approximately described by the coefficient N_P.

Due to this fact, the use of the first strategy is profitable only if $m_F \leq m/2$, i.e. if $N_P \leq 2$. If $m_F > m/2$, the cold redundancy (i.e. the use of reserve motor, for example) is more profitable than the application of PFCS1.

PFCS2 (PPM-based control strategy). If the PPM-based control strategy is used for the increase of time to the total failure (i.e. for the increase of reliability) of the system. One of the key elements of this strategy is the use of specific inverter control algorithms for the decrease of the motor torque oscillation magnitude in abnormal (i.e. post-fault) situations [5, 8].

2.6.2 Performance Evaluation of the Proposed Strategies

The abovementioned PFCS1 was selected as a base version of the corresponding system creation strategy for performance evaluation of the worked out PPM-based development and control strategy of fault-tolerant inverter-fed multiphase electric drive systems.

For performance evaluation of the PPM-based strategy of development of fault-tolerant inverter-fed multiphase electric drives, which is called "PPM-based strategy" below, the parameters $K_I \approx 1/H^{0.5}$ and $K_U \approx H^{0.5}$ were determined, where the parameter K_I shows in what times the electric motor phase currents may be increased when the PPM-based strategy is used in comparison with the case when PFCS1 is used, and the parameter K_U shows in what times the phase voltages must be increased when the PPM-based strategy is used in comparison with the case when PFCS1 is used.

The parameters K_I and K_U were determined for the case when in abnormal (i.e. post-fault) situation the torque created by the stator winding set, and speed of the motor secondary element should be the same as well as they were in normal (i.e. non-fault) situation.

The parameters K_I and K_U allows to evaluate approximately the change of mass and overall dimensions and manufacturing cost of the motor and inverter when worked out PPM-based strategy is used in comparison with the case when PFCS1 is used:

- the mass, overall dimensions, and manufacturing cost of the induction motor can be decreased approximately by a factor of $H^{0.5}$ (but no more this value) in passing from PFCS1 to the PPM-based strategy;
- the mass, overall dimensions, and manufacturing cost of inverter must be increased approximately by a factor of $H^{0.5}$ (but no more this value) in passing from PFCS1 to the PPM-based strategy.

For example, if $m = 6$ and PFCS1 is used, and the system must operate till the moment that $m_{F.\max} = 3$ phases will be damaged to, the time $T_O(m, m_F)$ is 1.5 times greater than the time $T_{O.3}$ of the 3-phase drive system operating to total failure (i.e. $K_{T_O} = 1.5$). However, in this case, the mass and overall dimensions and manufacturing cost of the whole inverter-fed electric drive and each of its elements will rise approximately by a factor of $N_P = 2$ (in comparison with the 3-phase system without any post-fault control strategy). Ultimately, the change in the mass and overall dimensions and manufacturing cost after the phase number increase and application of PPM-based strategy is as follows:

- the mass, overall dimensions, and manufacturing cost of the induction motor can be increased approximately by a factor of 2 (in comparison with the 3-phase system without any post-fault control strategy),
- the mass, overall dimensions, and manufacturing cost of inverter must be increased approximately by a factor of 2 (in comparison with the 3-phase system without any post-fault control strategy).

If passing from PFCS1 to the PPM-based strategy with $H = 2$, it is possible to decrease mass, overall dimensions, and manufacturing cost of the whole drive approximately by a factor of $H^{0.5} = 2^{0.5} \approx 1.4$ (however in this case the mass, overall dimensions, and manufacturing cost of the inverter will rise approximately by a factor of 1.4). Ultimately, the change in the mass, overall dimensions, and manufacturing cost after the phase number increase and application of PPM-based strategy is as follows:

- the mass, overall dimensions, and manufacturing cost of the induction motor can be increased approximately by a factor of 1.4 (in comparison with the 3-phase system without any post-fault control strategy),
- the mass, overall dimensions, and manufacturing cost of inverter must be increased approximately by a factor of 2.8 (in comparison with the 3-phase system without any post-fault control strategy).

Hereby the use of worked out PPM-based development and control strategy has advantages over existing control strategies. This strategy allows to develop the multiphase inverter-fed non-traditional controlled fault-tolerant electric drives with high reliability, less mass, and overall dimensions of the induction motor.

For a number of technical systems, in which the system reliability increase and the induction motor mass and overall dimensions decrease are very important (for example, oil deep-well pumps), the worked out PPM-based strategy is more preferable than PFCS1.

There is the following paradox in the field of the abovementioned systems. On the one hand, the number of wires between the inverter and motor is increased when the phase number m is extended. It must cause the system's reliability to decrease. However, on the other hand, system reliability greatly increases when the phase number m is extended owing to the possibility of PPM application in abnormal (i.e. post-fault) situations.

2.7 Conclusion

In order to fulfill the laws of energy efficiency [16, 17] operating in the field of multiphase asynchronous electric drives with more than four phases (in particular, when using phase-pole control in the drive system), the development of special designs of multiphase induction motors is required.

This chapter developed three basic variants of such structures: a motor with a toroidal multiphase stator winding (D1), a motor with a diamagnetic shield (screen) on the stator (D2), and a motor with a multi-rod stator winding (D3). Also developed are the principles for constructing power supplies for motors of the D3 type.

Further development of the concept of an induction motor with a multi-rod-type stator winding (D3) allowed the authors of the paper to develop principles for constructing windingless AC machines, which will have a simpler design and lower manufacturing cost than existing AC machines.

An increase in the number of phases and the use of phase-pole control method makes it possible to improve a number of technical and economic characteristics of induction motor drive systems, including increasing the reliability of the drive through the use of a post-fault control strategy based on the use of phase-pole control method, which has some advantages over other existing similar strategies.

References

1. I. Bolvashenkov, J. Kammermann, H.-G. Herzog et al., Design features and benefits of advanced multiphase inverter-fed electric drives, in *Proceedings of the 2nd International Conference on Sustainable Mobility Applications, Renewables and Technology "SMART 2022"*, November 23–25, 2022, Italy (2022)
2. E.A. Klingshirn, High phase order induction motors. IEEE Trans. Power Apparatus. Syst. **102**(1), 47–59 (1983)
3. P. Ferraris, M. Lazzari, Phase numbers and their related effects on the characteristics of inverter fed induction motor drives, in *Proceedings of the IEEE Industry Applications Society Annual Meeting IAS*, October, 3–7, 1983, Mexico City, Mexico (1983)
4. M.J. Duran, F. Barrero, Recent advances in the design modeling and control of multiphase machines—Part II. IEEE Trans. Industr. Electron. **63**(1), 459–468 (2016). https://doi.org/10.1109/TIE.2015.2448211
5. I. Bolvashenkov, J. Kammermann, H.-G. Herzog et al., Advanced control method for traction electric drives with multiphase induction motors: design and potential, in *Proceedings of the IEEE 14th International Conference on Ecological Vehicles and Renewable Energies "EVER'19"*, May 8–10, 2019, Monaco (2019), pp. 1–6
6. A.V. Brazhnikov, I.R. Belozerov, Prospects for the use of multiphase inverter-fed asynchronous drives in the field of traction systems of railway vehicles. Int. J. Railway **5**(1), 38–47 (2012)
7. A. Brazhnikov, N. Dovzhenko, A. Minkin et al., Novel type of EV hybrid traction drives. Int. J. Control Autom. **7**(3), 251–266 (2014)
8. A. Brazhnikov, N. Dovzhenko, Control potentials and advantages of multiphase AC drives, in *Proceedings of the 29th Annual IEEE Power Electronics Specialists Conference "PESC '98"*, Fukuoka, Japan, May 17–22, 1998, vol. 2 (1998), pp. 2108–2114

9. Osama, T.A. Lipo, Experimental and finite element analysis of an electronic pole-change drive. Paper presented at the IEEE-IAS Conference Record, Phoenix, AZ, October 3–7, vol. 2 (1999), pp. 914–921
10. A. Brazhnikov, Novel types of AC motors and drives for electrical and hybrid vehicles. J. Electr. Electron. Eng. **5**(1), 13–22 (2017)
11. I. Bolvashenkov, H.-G. Herzog, F. Ismagilov et al., *Fault-Tolerant Traction Electric Drives: Reliability, Topologies and Components Design* (Springer, Singapore, 2020)
12. I. Frenkel, L. Khvatskin, E. Ikar et al., Reliability and fault tolerance modeling of multiphase traction electric motors, in *Modeling and Simulation Based Analysis in Reliability Engineering*. ed. by M. Ram (CRC Press, Boca Raton, 2018), pp.265–293
13. S.N. Yeh, J.C. Hwang, T.T. Wu et al., Post-fault control strategy for six-phase permanent-magnet synchronous motor drives, in *Proceedings of the 30th ROC National Conference on Energy*, November 28–29, 2009, Taoyuan, Taiwan
14. J. Kammermann, I. Bolvashenkov, H.-G. Herzog, Improvement of Reliability and Fault Tolerance of Traction Drives by Means of Multiphase Actuators, in *Proceedings of the 7th VDE/VDI Symposium on Drive Systems*, November 22–23, 2017 (2017), pp. 83–88
15. A.V. Brazhnikov, E.S. Brazhnikova, R.V. Bondarenko et al., in *Speed Stepwise Controlled Multiphase Induction Motor Drive*. Russian patent No RU 130165 U1 10.07.2013 (2013) (in Russian)
16. A.V. Brazhnikov, I.R. Belozerov, Space-temporal spectral relations and energy efficiency invariance laws acting in the field of inverter-fed multiphase AC drives, in *Proceedings of the IET 6th International Conference on Power Electronics, Machines and Drives "PEMD '2012'"*, Bristol, UK, March 27–29, 2012, vol. 2 (2012), pp. 1094–1099
17. A.V. Brazhnikov, E.S. Brazhnikova, Efficiency invariance laws and development of multiphase AC inverter drives, in *Proceedings of the 21st International Symposium on Power Electronics, Electrical Drives, Automation and Motion "SPEEDAM '2012"*, Sorrento, Italy, June 20–22, 2012 (2012), pp. 420–425
18. A.V. Brazhnikov, *Multiphase Induction Motor*. RU patent 127268 U1, 20.04.2013 (2013) (in Russian)
19. N. Tesla, *System of Transmission of Electrical Energy*. US Patent 645,576, 20.03.1900 (1990)
20. N. Tesla, *Apparatus for Transmission of Electrical Energy*. US Patent 649,621, 15.05.1900 (1990)
21. N. Tesla, *Art of Transmitting Electrical Energy through the Natural Medium*. US Patent 787,412, 18.04.1905 (1995)
22. L. Anderson, *Nikola Tesla on His Work with Alternating Currents and Their Application to Wireless Telegraphy, Telephony, and Transmission of Power* (Twenty First Century Books, Breckenridge, CO, 2002)
23. B. Rzhonsnitsky, *Nikola Tesla. The First Domestic Byography*. Yauza-Eksmo, Moscow (2009) (in Russian)
24. T.D. Chu, *High Frequency Breakdown Voltage*. Superconducting Super Sollider Laboratory, Dallas, TX (1992)
25. J. Park, I. Hennis, H.W. Hermann et al., Gas breakdown in an atmospheric pressure radio-frequency capasitive plasma source. J. Appl. Phys. **89**(1), 15–19 (2001)
26. D.M. Glukhov, *Simulation of Multiphase Induction Motors in Abnormal Situations*. PhD thesis, Tomsk Polytechnical University, Tomsk (2005) (in Russian)
27. V. Romanovskiy, M. Sjubaev, I. Bolvashenkov, Selection basic data of electrical machines for electrical propulsion systems, in *Bulletin of the State University of Maritime and River Fleet*, Sankt-Petersburg, Russia (2015), pp. 172–178 (in Russian)
28. I. Odnokopylov, K. Obraztsov, G. Odnokopylov et al., Novel fault-tolerant concept for linear induction motor drives, in *Proceedings of the 8th International Symposium on Linear Drives for Industry Applications "LDIA '2011"*, Eindhoven, the Netherlands, June 3–6, 2011, Paper No. 215 (2011)

Chapter 3
Practical Application of Electric Propulsion Systems for the Regular Flights to the Asteroid Psyche

Abstract The project of a space train "Asteroidplane" for the regular delivery of astronauts onto the surface of space bodies located in the asteroid belt is being considered. The asteroidplane is formed in the circumterrestrial orbit of the International Space Station (ISS) from three electric rockets of the ER-7 type with electric rocket engines of the MARS type, which working substance is nitrogen, and the source of electricity is a semiconductor solar battery made of gallium arsenide. A new design solution for the asteroidplane in the presence of two takeoff and landing capsules will allow regularly delivering 6 astronauts to the surface of any asteroid and searching for minerals there. The calculations have shown that the developed optimal trajectory of the asteroidplane movement between the orbit of the Earth and the orbit of the asteroid Psyche will allow us to reach the asteroid Psyche in 55 days at the maximum speed of the asteroidplane of 200 km/s.

Keywords Asteroid belt · Electric flight · Asteroidplane · Electric rocket · Space train · Asteroid Psyche

3.1 Introduction

The chapter is devoted to reaching and mastering objects of the solar system. The matter we are talking about is conducting expeditions to the region of the asteroid belt, which is located between the planets Mars and Jupiter at a distance of 2–2.8 astronomical unit (AU) from planet Earth.

In [1], a calculated study was conducted, and an expedition project was carried out, in which it was shown that the dwarf planet Ceres could become the next object of human penetration into the solar system after Mars. Preliminary studies have shown that the industrial development of the asteroid belt, which involves the extraction of minerals in the bowels of asteroids, can begin as early as in the twenty-first century.

The observed progress in the development of space technology and especially automatic devices, makes it possible to begin geological exploration activities in the asteroid belt. The possibility of mining various metals, including noble and rare earths, which are in deficit on Earth has the greatest interest.

Several studies have been published on the first steps of the mastering of the asteroid belt, the generalization of which was carried out in [2]. This work continues to explore the possibility of penetration into the solar system in order to use the resources located in it. The possibility and economic feasibility of mining such natural elements as gold, platinum, nickel, cobalt, osmium, palladium, rhodium, rhenium, ruthenium, and others on asteroids should be confirmed by special studies.

The feasibility of this kind of search, as is known, is based on the scientific hypothesis, according to which, due to the low magnitude of gravity on the most asteroids, geological differentiation did not occur, as on planet Earth. Therefore, minerals should be sought at a shallow depth of penetration. To conduct the necessary research directly on asteroids, scientific expeditions should be conducted.

Delivery of scientific expeditions to the surface of asteroids is an important scientific and technical task, the attempt to solve which is the goal of this work. Is it possible in the near future to carry out regular flights of astronauts to any asteroid? To answer this question will help the research carried out by the authors.

3.2 The Concept of the Expedition to Psyche

As a result of preliminary research, standard designs of space trains were developed for flights to the planets of the solar system which are nearest to the Sun—Mercury, Venus, and Mars.

In this work it is shown that the optimal technical solution for the design of the space train lies in the way creating a new electric rocket engine using the phenomenon of superconductivity, while the energy source for powering this electric motor should be a solar battery operating on gallium arsenide.

Development has led to the creation of a new design of the ER-7 electric rocket, which houses superconducting electric rocket engines, a collapsible solar battery made of gallium arsenide and a container tank with the working substance (for the electric motor), which is in liquid state.

The space train is assembled in Earth orbit by docking electric rockets ER-7 and a takeoff and landing capsule in which the crew cabin is located. Each component of the train is launched into Earth orbit with the help of known, being now in service carrier rocket with chemical rocket engines. In 2021, a study was conducted on the electric propulsion system of an electric rocket train for a flight to the planet Mars. A detailed description of this study examines the process of joint operation of a superconducting rocket electric motor and an on-board power source from a solar battery [3].

The present work puts forward a new requirement for a space train, which should increase its operating range to a distance of 1.3 AU beyond planet Mars.

The process of space flight conducting is convenient to show on the example of an expedition to the asteroid Psyche. The asteroid Psyche was not chosen by chance.

3.2 The Concept of the Expedition to Psyche

It is known that Psyche is a main-belt asteroid that belongs to the metal-rich spectral class M. It is one of the ten most massive asteroids in the main belt. It is also known that Psyche has an oblong shape with dimensions of 240 × 185 km. It is located in the asteroid belt at a distance of 3.0 AU from the Sun.

Psyche makes one revolution in 5 years, simultaneously rotating around its own axis with a period of 4.2 h. At the same time, the perihelion of the Psyche trajectory is 2.54, and the aphelion is 3.22 AU. The acceleration of free fall on Psyche is 0.06 m/s^2, and the second space velocity is 0.154 km/s. This circumstance greatly facilitates the tasks of moving of the takeoff and landing capsule (TLC) above its surface.

At the same time, it should be borne in mind that the TLC can fall off the surface of the asteroid due to very little gravity. Therefore, there is a problem of fixing the TLC. Is it possible to find a solution to this problem with the help of harpoons and special rocket engines pressing the capsule against the surface of the asteroid?

It should also be remembered that due to the increase in the distance to the Sun, the power of the energy source—the solar battery—decreases by 7 times compared to the power in the Earth orbit. The question arises about the possibility of using a solar battery to power the rocket engines of a space train with the expedition crew.

The performed project of the flight of a space train to Psyche has showed that its flight trajectory to the asteroid belt consists of three sections, in each of which the gravitational field differs in magnitude.

On the first section of the trajectory: the surface of the Earth—the orbit of the Earth, the space train overcomes the gravitational force of the Earth. In this area, chemical rocket engines should be used, which are installed on the serially produced carrier rockets. On the second section of the trajectory: the orbit of the Earth—the orbit of the asteroid, the flight of a space train occurs in conditions of weightlessness. The use of chemical rocket engines in this case is not advisable.

The specific impulse of a chemical rocket engine is 400 s, whereas an electric rocket engine has a specific impulse of 8000 s. To reduce the consumption of the working substance of the rocket engine during the orbital flight, electric rocket engines should be installed on the space train.

On the third section of the trajectory, when the space train remains at the orbit of the asteroid, and the spacecraft takeoff and landing capsule with astronauts on board is under impact of the gravitational forces of the asteroid, a chemical rocket engine installed on TLC should be used.

The performed design and development works have shown that in order to solve the problem of regular delivery of astronauts to the surface of an asteroid with the minimum flight time, it is advisable to divide the process into two stages.

At the first stage, with the help of a carrier rocket equipped with a chemical rocket engine, the astronauts are delivered onto the international space station ISS. For this purpose, the project uses the Falcon 9 carrier rocket developed by Space X. For the flight on the second stage, the space train "Asteroidplane" was developed, delivering astronauts from the orbit of the planet Earth to the orbit of any asteroid, for example, to the orbit of the asteroid Psyche.

A general view of the "Asteroidplane" is shown in Fig. 3.1.

Fig. 3.1 General view of the "Asteroidplane"

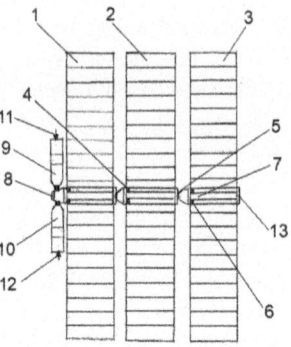

The design of the "Asteroidplane" is a system consisting of three interconnected electric rockets of the ER-7 type.

The forward electric rocket 1 is connected to the middle electric rocket 2 by the docking unit 4. The rear electric rocket 3 is connected to the middle electric rocket by docking unit 5. Each ER-7 rocket is equipped with 4 superconducting SERPS-type electric rocket engines, described in detail in [3]. The source of electricity for the operation of the engines is a solar panel made of gallium arsenide.

The working substance for superconducting electric motors is nitrogen, which in a liquid state is in the cryogenic tank 7 located along the longitudinal axis of the rocket.

The main design feature of the space train "asteroidplane" is that in the nose of the front electric rocket 1 there is a rotating shaft with a cylindrical nozzle 8 at the end. The cylindrical nozzle 8 is made with two external docking units located along the transverse axis.

With the help of these docking units, two takeoff and landing capsules 9 and 10, in which the astronaut cabins are located, are docked. When a space train reaches the orbit of the asteroid on which the expedition is supposed to be conducted, the space train is divided. The takeoff and landing capsules 9 and 10 are separated from the space train and make by turns landings on a pre-selected section of the asteroid surface.

For landing and takeoff of capsules 9 and 10, chemical rocket engines 11 and 12 are installed on them.

For repeated flights of the asteroidplane, the filling of tanks 7 with liquid nitrogen is carried out in Earth orbit by the space refueler using docking units 4, 5, and 13. A detailed description of the structural units of the electric rocket ER-7 is given in [3].

The expedition on the asteroidplane begins with the assembly of a space train in the orbit of the ISS, which is carried out automatically. With the help of the carrier rocket "Delta Heavy", the front orbital module 1 is launched into the Earth orbit. It is followed by the launching of middle module 2 and rear module 3. With the help of electric rocket engines 6, the approaching and docking rockets 1, 2, and 3 occur.

The takeoff and landing capsules 9 and 10 are launched into the circumterrestrial orbit of the ISS and are moored to the international space station. With the help of

3.2 The Concept of the Expedition to Psyche

a permanent system for delivering astronauts to the ISS, expedition members are taken to the ISS, where they use berths to pass into the cabins of TLC-1 and TLC-2. The assembly of the space train continues. The takeoff and landing capsules TLC-1 and TLC-2 are disconnected from the berths of the ISS. With the help of a chemical rocket engine 11 (Fig. 3.1), TLC-1 begins to move toward approaching with the space train from the side of docking unit 8 in the nose of rocket 1.

Next, the TLC-1 is docked with rocket 1. After that, the same operation is carried out with the TLC-2 on the opposite side using the chemical rocket engine 12. As the result of the approaching and the docking, TLC-2 is connected to docking unit 8 from the opposite side using a chemical rocket engine. The assembly of the space train asteroidoplane is finished. The artificial gravity system in the cabins of TLC-1 and TLC-2 is turned on. The shaft on which the docking station 8 is located (Fig. 3.1) begins to rotate in a superconducting bearing at a frequency of 5 rpm.

The detailed description of the system for creating artificial gravity is given below.

The second stage of the expedition begins—the movement of the asteroidplane from the orbit of the planet Earth to the orbit of the asteroid Psyche. To conduct the expedition, the trajectory of the asteroid was calculated using an astrodynamic program. The astrodynamic program continuously determines the force of gravity acting on the asteroidplane in the space between orbits of Earth and Psyche.

At the same time, at each moment of time, the force action of the Sun, Earth, Moon, Venus, and Mars and the largest asteroids is taken into account when their mutual position in the solar system changes [4]. The asteroidplane is a body of variable mass, since its electric rocket engines continuously eject working substance—nitrogen. At the same time, during the movement of the asteroidplane, there is a continuous decrease in the thrust force of electric rocket engines due to decrease in the power of solar panels by moving off from the Sun.

The calculation program takes into account the change in the thrust force of the engines using the method of successive approximations. The first flight on an asteroidplane from Earth's orbit to Psyche's orbit begins at point 1, shown in Fig. 3.2. At this moment, the asteroid Psyche is in its orbit at point $1'$.

Figure 3.2 shows the results of the calculation of optimal trajectories (1–8), which are proposed for the flight of the asteroidplane during 3 years in order to conduct researches and to create a scientific laboratory at Psyche with a permanent stay and change of crew of the expedition.

The results of the calculation of the change in the basic parameters of the asteroidplane in time during its movement between orbits are shown in Fig. 3.3.

As it can be seen from Fig. 3.3, in one day the speed of the asteroidplane reaches 10 km/s and it leaves the Earth's orbit.

Herewith the power of solar panels is 4.5 MW.

The asteroidplane gradually gains speed and after 22 days flies at a speed of 200 km/s. To carry out the movement of the asteroidplane to Psyche, a calculation program was developed, to select the optimal parameters of the flight. The optimization criterion is to achieve the orbit of the asteroid in the minimum time of the orbital flight. When the asteroidplane reaches its maximum speed, the power of the solar panels is reduced to a value of 3.0 MW. The mass of the space train is reduced from 72

Fig. 3.2 Results of calculation of optimal flight trajectories of the asteroidplane during 3 years

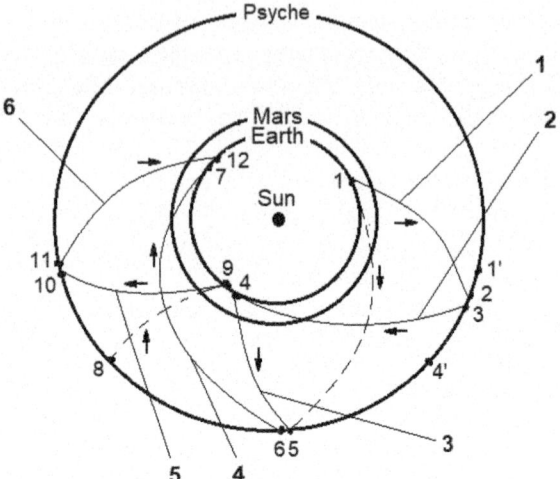

to 65 tons. After that, the electric rocket engines are turned off and the asteroidplane goes into inertia driving mode for 13 days.

After 35 days past the start of the movement, the asteroidplane makes 180° turn. Then the electric rocket engines 6 (Fig. 3.1) are switched on. The asteroidplane continues to move along the calculated trajectory in braking mode. Herewith at the end of the path, the power of solar panels that power electric rocket engines is reduced to a value of 2.1 MW.

In the braking mode, the asteroidplane is for 20 days. Herewith its speed decreases from 200 to 0.15 km/s. When reaching the orbit of Psyche, the asteroidplane flies around it at a distance of 30 km at a speed of 500 m/s.

The asteroidplane becomes a satellite of Psyche. It is now at point 2 (Fig. 3.2). One revolution around Psyche the asteroidplane makes in 110 min.

With the help of electric rocket engines 6 (Fig. 3.1), the asteroidplane flies around Psyche in order to reach the intended landing site of the expedition on its surface. With the help of a chemical rocket engine 11, the takeoff and landing capsule 9 TLC-1 (Fig. 3.1) is separated from the docking station 8 and makes a maneuver with access to the landing site on the surface of Psyche. Under the action of rocket engine 11 TLC-1 (takeoff and landing capsule 9) makes a soft landing. To prevent capsule 9 from bouncing off in contact with the asteroid surface, it is equipped with a special device actuated at the time of landing. A description of the design of the device is given below.

At the moment of TLC-1 landing on the surface of Psyche, the asteroidplane with the second capsule TLC-2 (capsule 10) continues to fly around Psyche, moving along its orbit.

When this revolution ends, the TLC-2 disconnects from the docking unit 8 (Fig. 3.1) TLC-2 and goes to landing, which is carried out with the help of a chemical rocket engine 12 at a distance of 50 m from the landing site of the first takeoff and landing capsule TLC-1. As it can be seen from Fig. 3.3, the flight from the Earth's

3.2 The Concept of the Expedition to Psyche

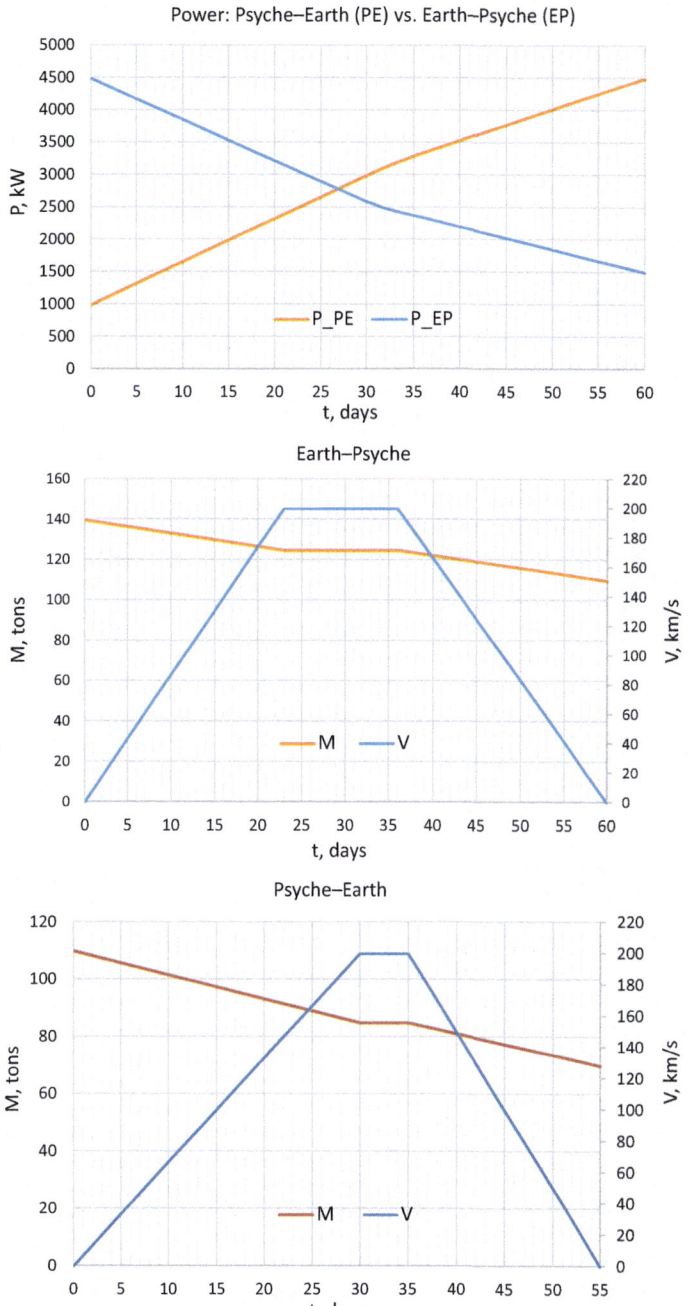

Fig. 3.3 Change in the basic parameters of the asteroidplane in time during its flight between orbits

orbit to the orbit of Psyche takes 58 days. The expedition begins its work on the surface of the asteroid according to the planned program. First of all, it is necessary to conduct an examination of the surface of Psyche, which has a complex geometric shape.

But it all starts from the moment when astronauts leave the takeoff and landing capsule and in light spacesuits go to the surface of the asteroid. There is a process of adaptation when astronauts move on the surface of the asteroid almost in conditions of weightlessness.

Clarification of existing geographical descriptions is carried out using a flyby of the surface of the asteroid Psyche on the takeoff and landing capsules TLC-1 and TLC-2, which serve as excellent rocket planes. It should be noted that the acceleration of free fall on the surface of Psyche is 160 times lower than on Earth. With the help of a chemical rocket engine installed in the bottom of the takeoff and landing capsule, the capsule rises above the surface of Psyche. Thanks to the steering device TLC is able to move to any point on the surface of Psyche at a speed of 200 km/h, brake, and make a landing and takeoff.

According to the expedition plan, after landing in the selected crater on the surface of Psyche, the astronauts leave the TLC and begin to conduct a soil study with the help of a drilling rig. An astronaut-geologist collects a collection of metals for delivery to Earth. It should be recalled that geological exploration is carried out in conditions of weightlessness at a surface temperature of -113 °C.

The daytime on the surface of Psyche is 126 min. According to the expedition program, the astronauts should build a permanent geological exploration laboratory and the first enterprise for the extraction of precious metals and rare earth elements on Psyche. Such a laboratory should eventually be expanded and turned into a space station on which it is possible to carry out work with a constant change in the composition of the expedition.

The first expedition lasting 10 days is of a reconnaissance nature. Astronauts conduct research, constantly staying in the cabins of the takeoff and landing capsules TLC-1 and TLC-2. After completing the expedition program, the astronauts make the takeoff of the TLC-1 and TLC-2 from the surface of Psyche and dock with the space train using docking unit 8 (Fig. 3.1).

The asteroidplane after assembling in the orbit of Psyche leaves its orbit around the Sun and takes a course into the orbit of the Earth. As it's shown at Fig. 3.2, while being in the orbit of Psyche for 10 days, the asteroidplane moves in its orbit around the Sun from point 2 to point 3. Further, the movement of the asteroidplane occurs along its trajectory 2 to point 4 into the Earth's orbit.

The calculation results of the parameters of the movement of asteroidplane in time along the trajectory 2 are shown in Fig. 3.3. Considering this process, we can see that within 30 days the asteroidplane is accelerated to a speed of 215 km/s. Herewith when approaching the Sun, the power of solar panels increases to a value of 2.8 MW.

After 5 days of flight by inertia after the maneuver of the asteroidplane, the direction of the nozzles of the electric rocket engines changes by 180°. The deceleration of the asteroidplane begins. Herewith the speed reduction from 215 to 10 km/s occurs in 12 days.

3.2 The Concept of the Expedition to Psyche

As it can be seen from Fig. 3.3, the flight from the orbit of Psyche to the orbit of the Earth, which takes place according to trajectory 2, takes 50 days. After the asteroidplane enters the Earth's orbit, first the TLC-1 takeoff and landing capsule, and then the TLC-2 with the help of docking unit 8 (Fig. 3.1) are disconnected from the asteroidplane and, under the action of chemical rocket engines 11 and 12, are sent to the orbit of the ISS. The capsules are taxied to the berths of the ISS station. Expedition crews move from the capsules to the station. This concludes the first reconnaissance expedition to Psyche, which lasted 123 days.

Let's hope that the successful results of the search for minerals of the first expedition will raise the question of conducting a cycle of further expeditions to Psyche and creating a base there with a permanent stay of astronauts. It should be noted that the Earth revolves around the Sun at an angular velocity 5 times faster than Psyche. Therefore, a second expedition to Psyche with the help of an asteroidplane should be carried out at the moment when the opposition of the Earth and Psyche occurs.

As it can be seen from Fig. 3.2, the trajectory of the first expedition ends at the point 4 of the Earth's orbit. At this moment, Psyche is in its orbit around the Sun and will be at point 4. Calculations show that the opposition of the Earth and the Psyche will occur 12 months after the end of the first expedition. It is then that the second expedition to Psyche should begin. The calculated flight trajectory of the asteroidplane from the Earth's orbit to the orbit of Psyche during the second expedition 3 (Fig. 3.2) passes from the point 4 to the point 5.

A one-year break between expeditions is used to study samples delivered from the surface of the asteroid. In order to conduct expeditions to Psyche, it is necessary to provide comfortable conditions for the work of the expedition crew with a constant change of crew. For this purpose on the Earth-Psyche route besides the asteroidplane it is necessary to carry out the movement of a freight electric train. The electric freight train is shown in Fig. 3.4.

It consists of two rockets 1 and 2 of type ER-7 and three takeoff and landing capsules 3, 4, and 5. The assembly of a freight electric train is carried out in Earth

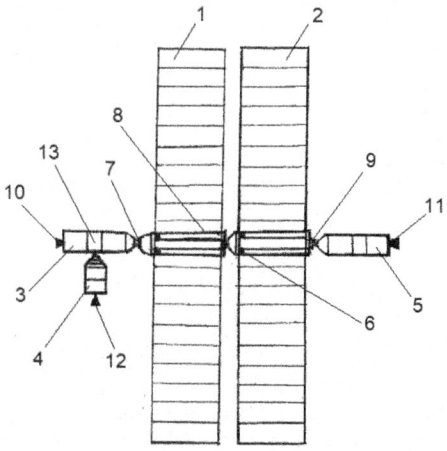

Fig. 3.4 Electric freight train

orbit at an altitude of 500 km from the Earth surface. Each element of the electric cargo train is put into orbit using a reusable Falcon rocket. Further, in the Earth's orbit, the assembly of a freight electric train is carried out according to the design scheme shown in Fig. 3.4. Takeoff and landing capsules 3 and 4 are interconnected using chemical rocket engines 10 and 12 operating in automatic mode.

Then using electric rocket engine 6 and docking unit 8 the rocket 1 is joined to rocket 2. With the help of a chemical rocket engine 10 and the docking unit 7, a device consisting of two takeoff and landing capsules 3 and 4 is assembled, which will later become the main link in the flying laboratory on Psyche, in which artificial gravity can be provided for astronauts. A detailed description of this device is provided below.

The assembly of the electric freight train ends after it is attached (using the docking unit 9 and the chemical rocket engine 11) to the takeoff and landing capsule 5, which serves for the regular delivery of goods between the orbits of Psyche and the Earth.

The calculations performed show that the launch of the freight train (Fig. 3.4) should be carried out 120 days before the start of the second expedition to Psyche. The trajectory of motion of the electric freight train 5 runs from point 1 in Earth orbit to point 5 in the orbit of Psyche. The second expedition to Psyche is conducted using an asteroidplane (Fig. 3.1) one year after the first expedition from point 4 in Earth's orbit along trajectory 3 to point 5 in the orbit of Psyche.

The task of the second expedition is to continue the study of the surface of Psyche and create a permanent laboratory there. The composition of the expedition is 6 astronauts, three of whom are placed in the TLC-1 capsule and three in the TLC-2 capsule. Upon arrival on the surface of Psyche, the astronauts assemble a flying laboratory consisting of two landing capsules, which were delivered by the electric freight train. The flying laboratory FL will allow not only to move on the surface of Psyche, but also provides for astronauts the possibility to constantly stay under the influence of the force of artificial gravity, which is created as a result of the rotation of capsule 3 relative to capsule 4 (Fig. 3.4). After assembling the flying laboratory, the astronauts install instruments and special equipment to study the surface composition of Psyche, which was delivered by block 5 of the cargo electric train shown in Fig. 3.4.

The stay of the expedition in its full staff lasts 20 days. After that, three astronauts take their places in the takeoff and landing capsule, and one pilot-astronaut takes a place by the control panel of the TLC-2 capsule. After the docking of the takeoff and landing capsules 9 and 10 (Fig. 3.1) with the asteroidplane, the electric rocket engines are turned on and the asteroidplane leaves the orbit of Psyche. It begins a flight along trajectory 4 from point 6 to point 7 in Earth's orbit. The remaining three astronauts at Psyche continue their planned researches. The Earth flies away from Psyche for a whole year and the astronauts move to the block 3 of the flying laboratory, where they are accommodated with convenience in the presence of artificial gravity and protection from cosmic radiation.

As the calculations show, the asteroidplane with astronauts who have left Psyche accelerates to a speed of 210 km/s, after which braking occurs to the speed of 12 km/s and after 55 days the asteroidplane reaches the Earth's orbit at point 7 (Fig. 3.2). In 200 days after the start of the expedition onto Psyche, astronauts send a cargo

3.3 The Structure of the Asteroidplane

electric train to Earth orbit along trajectory 6, passing from point 8 to point 9. Having completed their program, the astronauts meet on Earth the shift of the expedition, flying along trajectory 7 from point 9 to point 10 in the orbit of Psyche. In the future, the planned system of the expeditions with a constant change of astronauts-researchers can be continued and the asteroidplane will fly along the trajectory 8 from the point 11 on the surface of Psyche to point 12 in the Earth's orbit.

To carry out each regular flight of the asteroidplane, it is necessary to refuel electric rocket engines with the working substance—liquid nitrogen—in Earth orbit. In addition, it is necessary to refuel chemical rocket engines of the TLC with fuel—liquid hydrogen and an oxidizer—liquid oxygen. The design of the refueler and the technology of refueling the asteroidplane was developed earlier during development of the Moonplane-spacecraft design for the regular delivery of astronauts to the surface of the Moon [5].

3.3 The Structure of the Asteroidplane

The design of the asteroidplane is shown in Fig. 3.5, the longitudinal section, and in Fig. 3.6, the cross-section.

Here

1. Front orbital module ER-7-1.
2. Middle orbital module ER-7-2.
3. Rear orbital module ER-7-3.
4. The first takeoff and landing capsule TLC-1.
5. The second takeoff and landing capsule TLC-2.
6. Housing of the front orbital module.

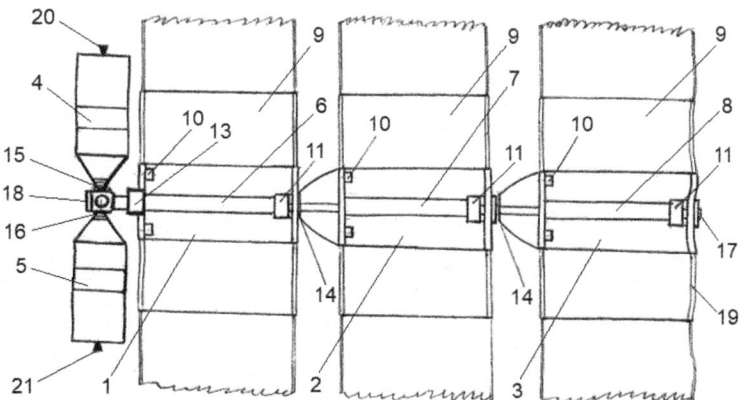

Fig. 3.5 Asteroidplane (the longitudinal section)

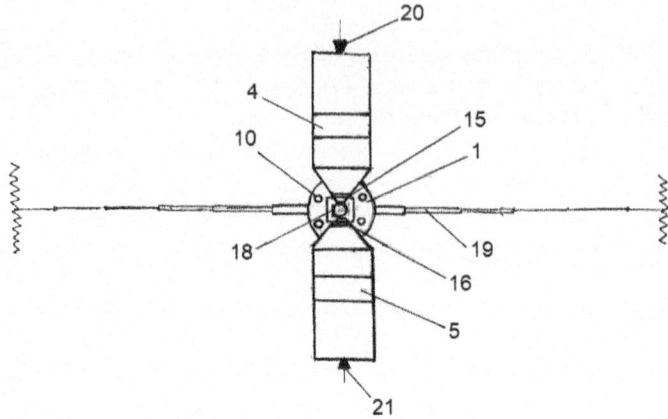

Fig. 3.6 Asteroidplane (the cross-section)

7. Housing of the middle orbital module.
8. Housing of the rear orbital module.

Inside the housings 6, 7, and 8 there are cryostat tanks with a working substance—liquid nitrogen for superconducting electric rocket engines 10, which are located in the end disks. The housings of electric rocket engines have the shape of a cylinder and are made of carbon fiber.

Outside, the cryostat tanks of electric motors have shield-vacuum insulation. The assembly of solar panels 9, which serve as the source of electrical energy to power electric motors, is made of gallium arsenide. In the end disks of the ER-7-1, ER-7-2, and ER-7-3 rockets there is a telescopic rods extension system 19, with the help of which a turn of the solar panels 9 passes along the transverse axis. Cryogenic pumps 11 are installed in the tail of the orbital modules to ill the cryostatic tanks with liquid nitrogen.

In the front end part of the module 1 a superconducting bearing 13 is installed which is used to rotate the shaft with the nozzle 18. The shaft nozzle 18 have two docking units 15 and 16, by which takeoff and landing capsules 4 and 5 dock with orbital module 1. The design of the asteroidplane will allow by rotating the takeoff and landing capsules of TLC-1 and TLC-2 relative to the main axis to create an artificial gravitational force field in the cabin of astronauts during an interorbital flight. In order the astronauts to feel comfortable during the flight (with gravity, as on Earth), the rotation speed of the capsules should be 5 rpm.

The connection of orbital modules 1, 2, and 3 made according to the structural scheme of the electric rocket ER-7 is carried out using docking units 14. To dock the asteroidplane with the space refueler during refueling with liquid nitrogen in Earth orbit, an orbital docking unit 17 is installed in the end part of orbital module 3.

Autonomous movement of the TLC-1 and TLC-2 during landing and takeoff from the surface of asteroids and the movement above their surface is carried out using chemical rocket engines 20 and 21. For the interorbital flights of the asteroidplane

3.4 Design of the Superconducting Electrorocket Motor MARS

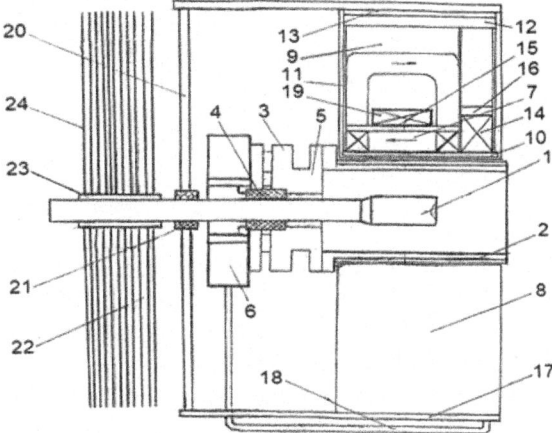

Fig. 3.7 Superconducting electrorocket motor MARS (longitudinal section)

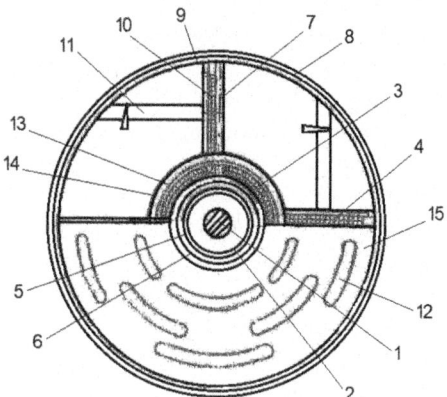

Fig. 3.8 Superconducting electrorocket motor MARS (cross-section)

and for its maneuvering in space, the superconducting electric rocket engines 10 are installed in the front end disks of the electric rockets ER-7-1, ER-7-2, and ER-7-3.

Figure 3.7 shows the longitudinal section and Fig. 3.8 shows the cross-section of the superconducting electric rocket motor 10.

3.4 Design of the Superconducting Electrorocket Motor MARS

For the presented project of the asteroidplane, the design of a superconducting electric rocket engine of the MARS type was developed [3]. The detailed calculated study was conducted, which showed that the developed design will create an electric rocket engine with high efficiency for space trains delivering astronauts participating in expeditions to the planets of the solar system [6].

The process of converting electrical energy into mechanical energy of the working substance ejected from the nozzle will take place in the working chamber, along the axis of which the cathode 1 is located, and on the outside—the anode 2 (Fig. 3.7), which has a cylindrical shape.

The cathode 1 is fixed to the anode 2 by a cylindrical bushing 3 and an insulator 4. The working substance of the electric motor is nitrogen with the addition of 0.2% cesium, which in a gaseous state at a temperature of 2500 K is fed into the interelectrode space. Nitrogen is supplied through the pipeline 18 to the chamber 6 of preparation of the working substance which has a ring shape.

Inside the chamber there is a dispenser and a solenoid valve. The anode 2 is installed and fixed in a cylindrical cavity of the cryostat 8, which is made of carbon fiber. Cryostat 8 houses the superconducting magnetic system of the electric motor. The superconducting magnetic system is made of a magnesium-boron compound. It consists of three windings for various purposes. The first winding 7 consists of four superconducting coils, with the help of which a tangentially directed magnetic field is created in the working chamber of the engine.

The second winding 14 is designed to create a magnetic field of radial direction. When the radial component of the magnetic field interacts with the current in the plasma, the working substance is compressed before being ejected into outer space (the so-called magnetic nozzle).

The third winding 19 is cylindrical in shape and is designed to stabilize the electrical discharge between electrodes 1 and 2 by uniform rotation of the electric arc. As it can be seen in Fig. 3.8, the coil of the first winding consists of two parts 3 and 4. The first part of the coil 3 with half-turns wound in the forward direction is placed on the outer surface of the cylinder 6.

Before laying this part of the coil 3 is shaped like a cylinder by forming. The other part of the coil 4 with half-turns wound in the opposite direction is bent and laid along the radius. Similarly, the other three coils of the first winding are laid. The coils of the first winding are secured by means of a cylindrical bandage 13 and a flat bandage 10, which presses the winding to the rib 7 by means of a winding holder 11. The inner cavity of the cryostat, which has an outer shell 9, is filled with the working substance of the engine—nitrogen, which is in liquid state at a temperature of 77 K.

The inner shell of the cryostat 5 is located outside anode 2 and is inserted inside the cylindrical body of the cryostat when the electric motor is assembled. The cylindrical outer housing, which in Fig. 3.7 is indicated by the index 17, has a disk 20 in the end part, which serves as a thermal bridge between the hot cathode 1 and the cold cryostat 8. The cathode 1 is fixed in the disk 20 using an insulating gasket 21.

In order to reduce the heat flow from the cathode to the cryostat in the disk 21 slots 12 (Fig. 3.8) are made, increasing the amount of thermal resistance.

Figure 3.7 shows the elements of intensive cooling of the cathode. For this purpose, a disk radiator 24 is installed on the surface of the cathode from the side opposite working chamber using a cylindrical hub 23, which gives off heat to the cosmic

3.5 Design of the Takeoff and Landing Capsule of the Asteroidplane

Table 3.1 Parameters of the electrorocket engine MARS

Parameter	Value
Thrust force	60 N
Power	600 kw
Current	600 A
Voltage	1000 V
Efficiency	94.5%
Working substance consumption	1 G/s
Magnetic induction	1.5 T
Anode diameter	170 mm
Cathode diameter	40 mm
Anode length	100 mm
Cathode length	60 mm
Outer cylinder diameter	650 mm
Outer cylinder length	450 mm

space. Reducing the surface temperature of the cathode reduces the amount of mass transfer of the cathode and increases the service life of the electric motor.

The calculated parameters and dimensions of the electric rocket engine for moving the asteroidplane from the Earth's orbit to the orbit of the asteroid Psyche are given in Table 3.1.

3.5 Design of the Takeoff and Landing Capsule of the Asteroidplane

The takeoff and landing capsule (Fig. 3.9) is designed to move three astronauts in orbit around the Earth and from the Earth's orbit to the orbit of any asteroid, for example, orbit of the asteroid Psyche.

Here

1. Chemical hydrogen–oxygen rocket engine with a thrust of 20 T and a specific impulse of 400 s.
2. The bottom of the rocket body of the capsule.
3. Tank-cryostat with liquid oxygen.
4. Expedition crew cabin.
5. Chemical rocket engine for maneuver.
6. Tank-cryostat with liquid hydrogen.
7. Outer shell (cylindrical part).
8. A superconducting solenoid for creation the magnetic field that protects astronauts against the flow of the charged particles in space.
9. Docking unit for connection to the shaft of the superconductor bearing.

Fig. 3.9 Design of the takeoff and landing capsule (TLC)

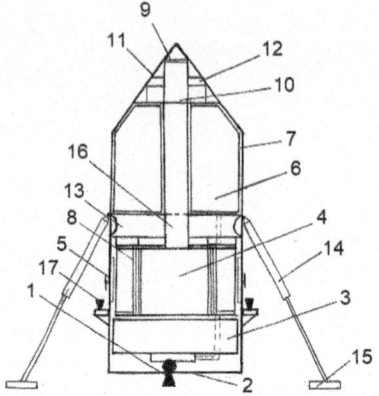

10. External tunnel and a gateway to exit the rocket.
11. The outer cone of the rocket housing.
12. Instrument container.
13. The power ring of the landing tripod.
14. Landing post damper.
15. Landing post shoe.
16. Internal tunnel and entrance into crew cabin.
17. Chemical oxygen–hydrogen rocket engine to press against the surface of the asteroid.

Mass of the TLC on the Earth—26 tons. TLC length—14 m. External diameter TLC—6 m.

3.6 The Design of the Superconductor Bearing for Creation of an Artificial Gravity System in the Astronaut Cabin

Artificial gravity in the cabin of astronauts in the takeoff and landing capsules TLC-1 and TLC-2 is created under the influence of centrifugal force acting when the capsules rotate relative to the longitudinal axis of the ER-7-1 rocket during orbital flights.

A new design solution was found that works on the Meissner effect and which is shown in Fig. 3.10, were

1. Rotating part.
2. Fixed part.
3. End clearance.
4. Protruding shaft.
5. A superconducting cylinder that rotates.
6. Outer shell of the yttrium-barium cylinder 5.

Fig. 3.10 Superconductor bearing for creation of an artificial gravity system in the astronaut cabin

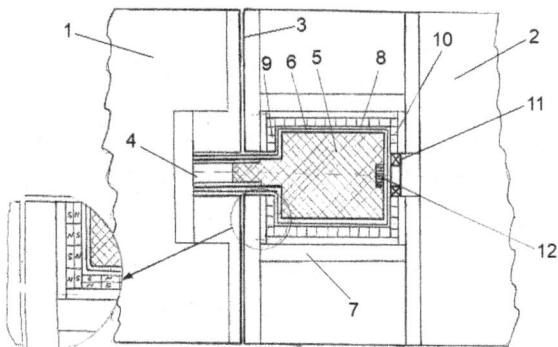

7. Bushing of the bearing.
8. Fixed cylinder with a set of permanent magnets which are made of iron–neodymium–boron compound.
9, 10. Sets of permanent magnets for axial force transmission.
11. Superconducting fixed electromagnet made of magnesium–boron compound.
12. Cylindrical ferromagnetic core of the superconducting electromagnet 11.

During the operation of the superconducting bearing, liquid nitrogen is fed into the gap between the rotating cylinder 5 and the fixed cylinder 8.

After cooling, cylinder 5 becomes superconducting and the magnetic field created by permanent magnets 8, 9, and 10 do not penetrate into the cylinder 5. Therewith, a lift arises that steadily holds the rotating part of the bearing 1 on the horizontal axis of the ER-7-1 rocket.

Rotor 5 is fixed by a fixed superconducting electromagnet 11 with a ferromagnetic core 12.

Thanks to magnetic interaction of the fixed and rotating part, the bearing shown in Fig. 3.10 has no energy losses for friction, which ensures minimal energy consumption for the creation of artificial gravity in the astronaut's cabin of the takeoff and landing capsule.

3.7 The Basic Module for Creating Artificial Gravity During the Stay of the Expedition on the Surface of the Asteroid

The basic modules for creating artificial gravity during the stay of the expedition on the surface of the asteroid is presented in Fig. 3.11, were

1. Rotating module with expedition crew.
2. A fixed module fixed on the surface of the asteroid.
3. Tank-cryostat with liquid oxygen.

Fig. 3.11 Basic module for creating artificial gravity

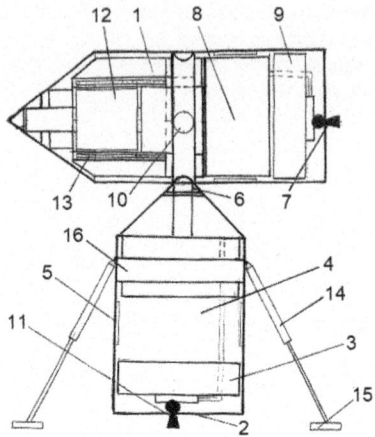

4. Instrument compartment for research on the surface of the asteroid.
5. Manhole for entry into the instrument compartment.
6. Docking unit with a superconducting bearing installed inside.
7. Chemical rocket engine of hydrogen–oxygen type.
8. Household compartment of the rotating module.
9. Tank-cryostat with liquid oxygen.
10. Window for asteroid surface observation.
11. Chemical rocket hydrogen–oxygen engine of fixed module.
12. Expedition crew cabin.
13. Superconducting solenoid protecting the cabin from the flow of charged particles.
14. Damper of the tripod post of the fixed module.
15. Shoe and harpoon of fixed module tripod post.
16. Support ring of the tripod post.

3.8 Conclusion

A complex of design and engineering works was carried out to create a space train—an asteroidplane for high-speed regular delivery of astronauts to the surface of space bodies located in the asteroid belt.

The design of the asteroidplane has been developed, which is formed in the Earth's orbit by docking individual modules delivered to the orbit using existing carrier rockets.

The movement from the Earth's orbit to the orbit of a pre-selected asteroid is carried out with the help of three interconnected electric rockets ER-7, in the bodies of which superconducting electric rocket engines of the MARS type are installed, the working substance for which is nitrogen, and the source of electro energy is a semiconductor solar battery made of gallium arsenide.

A new design solution of the asteroidplane in the presence of two takeoff and landing capsules will regularly be delivered to the surface of the asteroid 6 astronauts who are located in the cabin having a system of magnetic protection against cosmic radiation and a system of artificial gravity.

Design and development of the first exploratory research expedition to the asteroid Psyche was carried out in order to determine the possibility of mining minerals, mainly noble and rare earth metals in the bowels of the asteroid. The calculation of the optimal trajectory of the asteroidplane movement between the orbits of the Earth and Psyche has showed that the developed design of the asteroidplane allows us to reach the asteroid Psyche in 55 days at a maximum speed of 200 km/s.

References

1. A. Rubinraut, The expedition on Ceres. Int. J. Emerg. Technol. Adv. Eng. **8**(11), 1–13 (2018)
2. D. Lauretta, Asteroiden space. Bergbau **2**, 68 (2017)
3. I. Bolvashenkov, J. Kammermann, A. Rubinraut et al., *Vehicle Electrification: On Water, in Air, and Space* (Springer, Switzerland, 2022)
4. A.E. Roy, *Orbital Motion* (CRC Press, Boca Raton, 2020)
5. A. Rubinraut, Moonplane—a spacecraft of regular delivery of astronauts onto the moon. Adv. Aerosp. Sci. Technol. **4**(3), 43–56 (2019). https://doi.org/10.4236/aast.2019.43004
6. A. Rubinraut, The study of the electric rocket engine for the future. Adv. Aerosp. Sci. Technol. **2**(1), 1–16 (2017). https://doi.org/10.4236/aast.2017.21001

Chapter 4
Feasibility of Electric Propulsion System for the Regular Flights onto Titan

Abstract The project of the space train "Titanplane" for astronauts' regular delivery onto Titan—a satellite of the planet Saturn, is being considered. At the first stage of the flight, with the help of a carrier rocket equipped with a chemical rocket engine, astronauts are delivered to the International Space Station (ISS). To deliver astronauts from Earth's orbit to Titan's orbit, a space train has been designed consisting of six rockets equipped with superconducting electric and chemical rocket engines. For the first time for the movement of a space train, a superconducting electric rocket engine has been developed, equipped with three working chambers, the magnetic field in which is created by two opposite connected flat electromagnets wound from superconducting wire. To supply the electric motor with electric current, an on-board power plant has been developed, consisting of a gas-phase nuclear reactor, a superconducting magnetohydrodynamic (MHD) alternator and a cryoturbogenerator. The working substance for creating the jet thrust of an electric rocket engine—hydrogen is stored in liquid state in two cryogenic tanks—containers located along the longitudinal axis of the train. In the bow of the space train in the tank-container with the working substance there is a shaft rotating in a superconducting bearing. The shaft has a cylindrical nozzle, on which, with the help of docking units two takeoff and landing capsules are installed for landing and takeoff from the surface of Titan. When the shaft rotates in the capsules, a centrifugal force equal to the earth's gravity occurs. Refueling a space train with liquid hydrogen in Titan's orbit is carried out by a space refueler with a chemical rocket engine. The developed space train is capable of delivering six astronauts regularly to any point. When the shaft rotates in the capsules, a centrifugal force equal to the earth's gravity occurs. Refueling a space train with liquid hydrogen in Titan's orbit is carried out by a space refueler with a chemical rocket engine. The developed space train is capable for regularly delivering six astronauts to any point on the surface of Titan in 65 days.

Keywords Saturn · Titan · Superconductor · Magnetoplasma · Electric rocket motor

4.1 Introduction

The chapter is devoted to the problem of solar system mastering, which is carried out by the consistent conduct of expeditions to the planets and their satellites. This path was outlined by the authors in the projects of space trains developed in 2015–2020 to conduct expeditions to all the planets of the solar system. In accordance with the planned program, a project of an expedition to the planet Saturn was developed with the visit to the surface of its satellite Titan.

The design of a space train, where the locomotive of which has electric rocket engines with a superconducting excitation winding with a working chamber having a cylindrical shape was presented in [1]. In this case, the cathode of the working chamber is located along the axis of the engine, and the anode is installed outside. The working chamber is surrounded by a superconducting excitation winding, the current in which is directed along the axis of the motor. This makes it possible to create a magnetoplasma electric rocket engine with high efficiency and significantly reduce the mass of the on-board power plant of the space train.

It should be recalled that the study of the planet Saturn and its satellites has been conducted by astronomers for more than one year. The most significant results were obtained with the help of the Cassini spacecraft, which was an artificial satellite of Saturn during 2004–2017. During research, it approached Saturn at a distance of 950 km. And on December 25, 2004, the Huygens probe separated from Cassini, which descended on Titan on January 14, 2015. It turned out that at the landing site, Titan has a solid surface covered with water ice at a temperature of -179 °C.

This allows us to hope that the day will come when a human foot will set foot on the surface of Titan. Titan is known to be of particular interest to scientists. It is the second body after planet Earth to have a dense atmosphere of nitrogen. In addition, on its surface there are rivers and lakes of ethane and methane.

Of course, we should not forget that the distance from the orbit of the Earth to the orbit of the planet Saturn is 1280 million km. Therefore, there is the problem of delivering an expedition to the surface of Titan. The calculated study conducted in [1, 2] showed that the use of the latest scientific and technical technologies in the field of astronautics and, in particular, superconducting electric machines will allow a space expedition consisting of 3 people to reach the orbit of the planet Saturn and its satellite Titan in 70 days. The analysis of the developed project showed that the flight path to Titan consists of three sections, in each of which the gravitational field differs significantly in magnitude.

On the first section of the trajectory: the surface of the Earth—the orbit of the Earth, where the formation of a space train occurs by docking individual rockets, the space train overcomes the gravitational force of the Earth. On this section of the trajectory for the movement of the train should be used known launch vehicles with chemical rocket engines.

On the second section of the trajectory: the orbit of the Earth—the orbit of Titan, the flight of a space train occurs in conditions of weightlessness. The use of chemical rocket engines in this case is irrational. After all, the specific impulse of a chemical rocket engine is 400 s, while an electric rocket engine provides a specific impulse value of 12,000–15,000 s. To minimize the consumption of the working substance during the interorbital flight of the space train, an electric rocket engine must be installed on it.

On the third section of the trajectory, when the space train remains in orbit of Saturn, and the takeoff and landing capsule with astronauts on board are under the influence of Saturn's gravitational force, a chemical rocket engine should be used.

The completed design and engineering developments have shown that in order to solve the problem of regular delivery of astronauts to the surface of Titan with a minimum flight time, it is advisable to divide the process into three stages.

At the first stage of the flight, with the help of a launch vehicle equipped with a chemical rocket engine, astronauts are delivered to the International Space Station (ISS). For this purpose, the project uses the Atlas V launch vehicle developed by Boing with the CTS-10 Starliner space capsule.

For the second and third stages, a space train is being developed, which is by parts launched into Earth orbit with the help of known launch vehicles. Further, by docking, the train formation is carried out.

An electric rocket engine and two takeoff and landing capsules are installed on the interorbital space train. Such a design solution for the space train will allow for the regular movement of 8 astronauts from the ISS berth in Earth orbit to the orbit of Saturn with access to the orbit of Titan and landing on its surface.

After delivery to Titan, astronauts perform the planned program of work. At the third stage, a space train flies from the orbit of Titan to the orbit of the Earth. Developing space train project to regularly deliver astronauts to Titan is the goal of the present study.

4.2 The Concept of Delivering Astronauts to Titan Using the Space Train "Titanplane"

To deliver astronauts to Titan from the Earth orbit, in the project a space train called a "Titanplane", the general appearance of which is shown in Fig. 4.1, is developed.

The space train consists of five parts, each of which is launched using existing launch vehicles and docks with others in Earth orbit. The first part of the Titanplane—a space locomotive 1, which houses an electric rocket engine 6 and an on-board power plant.

The next two parts are tanks-containers 3 and 4, in which the working substance of the electric rocket engine 6 is in liquid form. Space locomotive 1 is connected to the container- tanks by docking units 13 and 5.

The space train has a design feature. In the bow of the tank-container 4 there is a rotating shaft 11 with a cylindrical nozzle 16 at the end. The cylindrical nozzle16 is made with two external docking units 14 and 15, located along the transverse axis of the Titanplane. With the help of these docking units the takeoff and landing capsules

Fig. 4.1 Space train "Titanplane"

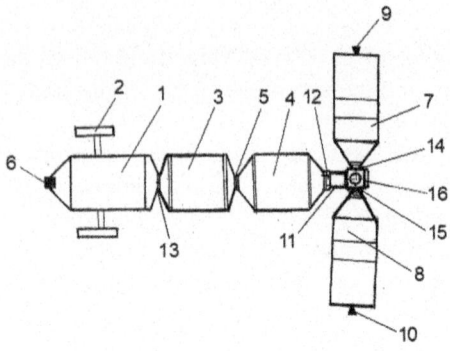

7 and 8 are docked. After the Titanplane reaches the planet Saturn and enters orbit of its satellite Titan the takeoff and landing capsules 7 and 8 with the help of chemical rocket engines 9 and 10 separate from the space train and land on the icy surface of Titan.

A detailed description of the design of the space locomotive and takeoff and landing capsule is given below. Let's follow how astronauts are delivered to the surface of Titan and their return to Earth's orbit.

It should be borne in mind that the regular delivery of astronauts to Titan should be carried out when a scientific research base (SRB), description of which is given in [1], is already functioning on its surface. The SRB is located on a flat plain on Titan's equator that has been dubbed Xanadu [3]. The coordinates of the SRB 37°/ 32° are determined from the geomorphological map of Titan, given in [3], based on the research of the Cassini.

The research base on Titan [3] is equipped with everything necessary to provide astronauts who work with measuring equipment under a special program. Already after the first expedition, a space refueler is on the ice surface, which produces hydrogen from water ice to refuel container tanks 3 and 4. The space refueler was developed earlier to provide an expedition to Jupiter [4].

The work [4] examines in detail the design of the space refueler and its functioning according to the developed technological scheme for the case when it is on the ice surface of the satellite of Jupiter—Europa. The space refueler allows not only to fill the tanks of a space train in orbit of Titan, but also to fill the tanks with liquid hydrogen and liquid oxygen located in takeoff and landing capsules 7 and 8 during their stay at the base.

The SRB is equipped with the installation of a special life support complex "Iglus" [5], which protects astronauts from cosmic radiation and the negative effects of the atmosphere and creates comfortable conditions during a long stay on the surface of Titan. The SRB has a device for moving along the ice surface of Titan, manufactured by RR. The developers are engineers at Londoner Royal College [6].

The work [1] also provides data on the possibility of using a hang-glider to move astronauts above the surface of Titan. After all, Titan has a dense atmosphere, and the force of gravity on its surface is 7 times lower than on the surface of the Earth.

4.2 The Concept of Delivering Astronauts to Titan Using the Space Train ...

The fuel for the internal combustion engine of the hang-glider is hydrogen, and the oxidizer is oxygen, which are available on board the takeoff and landing capsule.

The regular delivery of astronauts to the surface of Titan consists of several stages. The first stage—the movement of astronauts from the surface of the Earth to the surface of the international space station ISS, which is in orbit around the Earth—is carried out with the help of two Atlas V launch vehicles, in the head part of which the Orion spacecraft is installed. The group consists of 6 astronauts-researchers of various specialties and two astronauts-drivers, who will control the Titanplane during the interorbital flight. After moving to the ISS, astronauts are waiting for the beginning of the movement of the Titanplane.

The second stage of the flight from the Earth's orbit to the orbit of Titan begins with the assembly of the Titanplane in the orbit of the ISS, where the locomotive 1 of space train is first launched using the Delta Heavy carrier rocket After that, the tank-container 3 with a working substance for an electric rocket engine 6 is also launched into the ISS orbit with the help of the Delta Heavy carrier rocket. By the next launch with the help of the carrier rocket is a second tank-container 4 with an axial shaft 11 is launched.

Further, in automatic mode, with the help of an electric rocket engine 6, the locomotive 1 and the cars of space train 3 and 4 are docked using docking units 13 and 5. The launch of takeoff and landing capsules 7 and 8 into the orbit of the ISS is being carried out. For this purpose, the Delta Heavy carrier rocket is also used.

After entering the orbit of the ISS, space takeoff and landing capsules 7 and 8 are sent to the berths of the ISS using chemical rocket engines 9 and 10. After taxiing, the TLC-7 and TLC-8 are automatically docked with the ISS berths.

The crews of the Titanplane go to the cabin of the TLC-7 and to the cabin of the TLC-8. The assembly of the space train continues. The takeoff and landing capsule 7 with the help of a chemical rocket engine 9 begins to move towards convergence with the space train from the side of the docking unit 14 (Fig. 4.1). After docking the TLC-7 with unit 14, the operation of docking the second take-off and landing capsule 8 is carried out in using the chemical rocket engine 10.

As a result of convergence and docking the TLC-8 is connected to docking unit 15. In this way ends the assembly of the Titanplane in orbit around the Earth. The artificial gravity system in the cabins of the TLC-7 and TLC-8 is switched on. The shaft 11, on which the nozzle 16 with docking units 14 and 15 is located, begins to rotate in the superconducting bearing 12 with a rotational speed of 5 rpm.

The third stage begins—the movement of the space train from the Earth's orbit to the orbit of Titan. When designing the Titanplane, the calculation of the trajectory of its movement was performed.

The calculation was carried out using an asrodynamic program that determines the gravitational field in the space between the orbits of the planets Saturn and Earth when moving along the calculated trajectory. At the same time, at each moment of time, the force interaction of the Sun, Earth, Moon, Mars, the asteroid belt, Jupiter, and Saturn is taken into account during their joint movement in orbits [7].

Fig. 4.2 Trajectory of motion

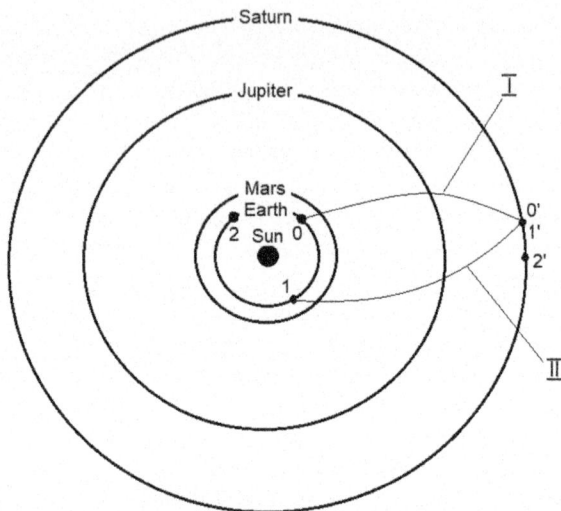

This determines the picture of the gravitational field of the solar system in the zone of motion of the space train, which is a body of variable mass, which is under the influence of this field.

The trajectory of motion obtained as a result of the calculation is shown on Fig. 4.2.

The process of changing the mass and speed of the Titanplane during the flight time is shown in Fig. 4.3.

The movement of the Titanplane begins after turning on the electric rocket engine 6. At the same time, the space train accelerates, increasing its speed of movement, and after reaching the second cosmic velocity, it enters the calculated flight path.

Calculations performed with the help of computer programs have shown that the flight path to Saturn is close to the trajectory along which the Kassini spacecraft flew in 2004 [8].

Figure 4.2, which shows the calculated trajectory of the Titanplane's flight, shows the location of the planetary orbits of the Earth, Mars, Jupiter and Saturn. At the beginning of the orbital flight (point 0), the Ttitanplane accelerates, moving along the trajectory 1 to point $0'$.

Further, during the two weeks stay in the orbit of Saturn, the Titanplane is moved from point $0'$ to point $1'$. After that, the Titanplane is sent from the orbit of Saturn to the orbit of the Earth. The motion of the Titanplane occurs according to the calculated trajectory II (Fig. 4.2) from point $1'$ to point 1. When the Titanplane reaches Earth's orbit, Saturn will already be at point $2'$. From Fig. 4.2, it can be seen that the next orbital flight to the orbit of Saturn should begin 11 months after the first flight, when the Titanplane will be at point 2 in Earth's orbit.

Let's trace with the help of Fig. 4.3 how the process of the Titanplane movement in time occurs. Electric rocket engine 6 during acceleration of the Titanplane creates a maximum thrust of 3000 N. At the acceleration stage, the working substance

4.2 The Concept of Delivering Astronauts to Titan Using the Space Train … 77

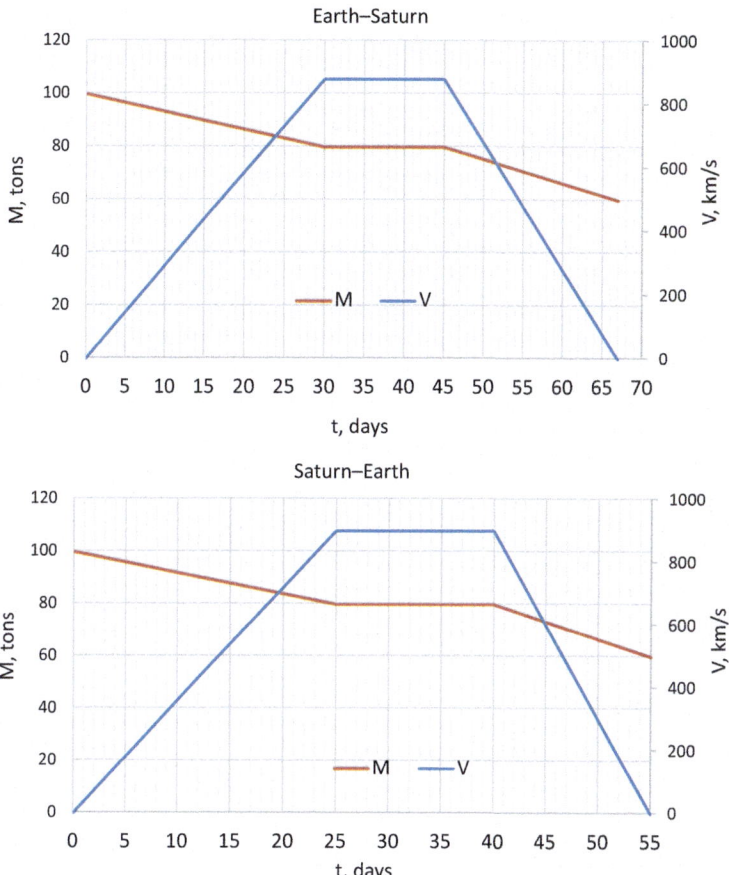

Fig. 4.3 Mass and speed of Titanplane

located in container tank 3 is consumed (Fig. 4.4) and the mass of the rocket train is continuously reduced. This leads to an increase in its acceleration.

After 29 days of flight, the speed of the Titanplane reaches 865 km/s. After the working substance in the tank-container 3 is consumed, the electric rocket engine 6 is turned off. The Titanplane continues to move along the calculated trajectory in free flight by inertia at a constant speed. With the help of the electric motor 6, only the

Fig. 4.4 Rockets connection

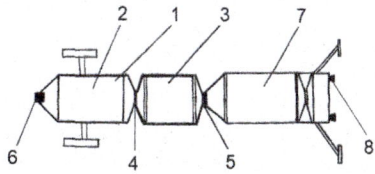

direction of movement is adjusted. The flight of the Titanplane at a constant speed lasts 16 days. After 16 days, the cruise rocket engine 6 turns on again. But at this time, the Titanplane makes a turn of 180° and the braking process begins. The speed of the Titanplane decreases and it approaches the planet Saturn at the same time, the working substance is consumed in the tank-container 4.

In 65 days after the start of the flight, the Titanplane reaches the orbit of Saturn. Titanplane with the help of an electric rocket engine 6 makes a maneuver and begins to move towards Saturn's "moon" Titan. It should be recalled that Titan is located behind the rings of Saturn, at a distance of 1,200,000 km from it Titan makes one revolution around its own axis in 16 days rotating symmetrically with Saturn.

After approaching Titan, the Titanplane reduces its speed and enters a circular orbit around Titan and becomes its satellite. The flight around Titan is carried out outside its atmosphere at 500 km from its surface at speed of 2.6 km/s. The landing of astronauts on the surface of Titan begins. With the help of rocket engine 6, the driver-astronaut brings the Titanplane above the surface in the region of the equator where the scientific research base (SRB) is located.

Now the Titanplane moves in orbit around Titan at the latitude of 15° at the speed of 1.9 km/s at an altitude from the surface of 420 km. When approaching the point with coordinates of 15°/30°, the astronaut-driver receives a signal from radio beacon on the territory of the SRB in the Xanadu desert [3], the surface of which is covered with water ice.

A command is underway to undock the Titanplane and the takeoff and landing capsule TLC-1 (7 in Fig. 4.1). Capsule 7 departs from the Titanplane. The driver-astronaut turns on the chemical rocket engine 9 and TLC-1 goes into braking mode. The Titanplane continues to move around Titan, and TLC-1 reduces its speed to a value of 1.6 km/s. At the same time, TLC-1 loses weightlessness and rushes to the surface of Titan. At an altitude of 100 km above the surface of Titan the TLC-1 parachute system is activated.

With the help of a chemical rocket engine, the driver-astronaut smoothly lands the capsule on the surface of Titan. After 20 min, the Titanplane again approaches the landing site. The second driver-astronaut, who controls the TLC-2, undocks unit 15 (Fig. 4.1), connecting the TLC-2 with the space train.

With the help of the chemical rocket engine 10 TLC-2 moves away from the Titanplane and moves in the mode of braking to the surface of Titan. With the help of a parachute system and a chemical rocket engine 10, the TLC-2 is smoothly landed on the surface of Titan in 200 m from the TLC take-off and landing capsule already located there.

Astronauts which arrived at the SRB are accommodated in the living quarters of the base and, after adaptation, begin their work according to the planned program. There are 9 astronauts-researchers permanently on Titan's SRB. The period of stay on the basis is 11 months. The transfer of the watch to the new crew of the expedition lasts 14 days. Astronauts who must leave the base and return to Earth have done the preparatory work.

With the help of the SRB refueler is prepared the necessary amount of working substance for the operation of chemical rocket engines to move from the surface

4.2 The Concept of Delivering Astronauts to Titan Using the Space Train ...

Fig. 4.5 Space train refueler

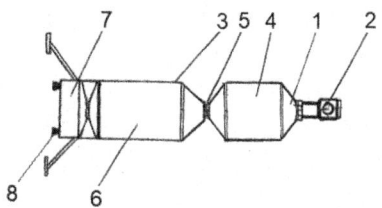

of Titan to the orbit around Titan, on which the Titanplane is located. These are hydrogen and the oxygen, which in the liquid state are stored in cryogenic tanks at short distance from the landing site of the TLC-1 and TLC-2.

After arrival, the astronaut team refuels the chemical rocket engines of capsules 7 and 8 using the technology described in [1]. After that, the team proceeds to refuel the working substance of the electric rocket engine 6 of Titanplane, filling container-tanks 3 and 4 with liquid hydrogen.

A diagram of the connection of the rockets for refueling tank-container 3 is shown in Fig. 4.4.

In automatic mode, with the help of docking unit 5, the Titanplane is divided into two parts. Space refueler 7, which has a chemical rocket engine 8, takes off from the surface of Titan and enters its orbit, where it docks with container-tank 3. With the help of a cryogenic pump, liquid hydrogen is pumped from the refueler tank 7 to the cryostat tank 3. After refueling, the space refueler 7 returns to base using a chemical rocket engine 8. For the first time, the process of filling tanks with the help of the space refueler was considered in [4], which describes the work of the refueler for an expedition to the planet Jupiter.

After refueling the cryostat tank on the locomotive side, the cryostat tank is refueled on the side of the shaft, on which the takeoff and landing capsules are mounted. A refueling diagram is shown in Fig. 4.5.

The space refueler 3, which has been on the surface of Titan, where its liquid hydrogen tank 6 was filled with a working substance for an electric rocket engine, and tank 7 is filled with liquid oxidizer (oxygen) for chemical rocket engine 8.

After that, refueler 3 takes off from the surface of Titan using a chemical rocket engine 8 and enters orbit around Titan. Refueler 3 approaches the container-tank 4 and docks to it by the unit 5. After pumping liquid hydrogen into the tank-container 4, the second half of the Titanplane is ready for docking. With the help of a chemical rocket engine 8 and an electric rocket engine 6 (Fig. 4.4), both halves of the Titanplane in orbit around Titan are docked. And the space refueler 7 with the help of a chemical rocket engine 8 returns to the zone of its permanent location on the surface of Titan, when transferring the watch to the new crew of the Titanplane. The team of astronauts, consisting of 10 people, leaves the research base of the Titan.

Departing astronauts are accommodated in the cabins of the TLC-1 and TLC-2. The chemical rocket engine 9 (Fig. 4.1) is switched on. TLC-1 takes off from the side of the SRB and after 80 s enters the orbit of Titan. Approaching and docking is carried out with docking unit 14 on the shaft 16, which is terminated in the superconducting bearing 12. After that, a similar operation is carried out with the TLC-2, using docking unit 15 (Fig. 4.1).

The interorbital flight of the Titanplane from the orbit of Titan to the orbit of the Earth begins. The system for creating artificial gravity in the cabins of astronauts staying in the cabins of the TLC-1 and the TLC-2 capsules is turned on.

After that, the astronaut-driver turns on the electric rocket engine 6. The speed of the Titanplane increases and reaches a value of 2.8 km/s. The Titanplane leaves the orbit of Titan and makes its way to the planet Saturn. Moving along the calculated trajectory with constant acceleration, the Titanplane in two days approaches the outer ring of Saturn (E ring). Further, the Titanplane continues to move along the radius of the rings of Saturn at a distance of 1000 km from the plane of the rings of Saturn. When approaching Saturn, the Titanplane makes a maneuver and enters a circular orbit around Saturn at a distance of 500 km from its surface. Titanplane becomes an artificial satellite of Saturn. After the flyby of Saturn, the Titanplane carries out the last stage of the expedition—a flight to the orbit of the planet Earth. During the flight in the Saturn zone, the Titanplane moves with it along its orbit from point $0'$ (Fig. 4.2) to point $1'$ at the speed of 9.7 km/s. Planet Earth, moving in its orbit, at this time moves from point 0 to point 1. And the Titanplane in order to return to Earth, must go along the calculated trajectory from point $1'$ to point 1.

The electric rocket engine 6 is switched on and the Ttitanplane accelerates. As can be seen from the graph of Fig. 4.3, after 26 days of flight, the speed of the train reaches 900 km/s.

The electric motor 6 is switched off. After a free inertial flight for 15 days, the electric motor 6 is turned on again, and the Titanplane is turned on 180°. The breaking of the Titanplane begins, which lasts 12 days. As it can be seen from the Fig. 4.3, the mass of the Titanplane is continuously decreasing over time due to ejection of the working substance during electric engine operation from 100 to 60 tons, which leads to an increase in acceleration. The duration of the flight of the Titanplane from the orbit of Saturn to the orbit of the Earth is 52 days. With decrease in speed to value of 8 km/s, the Titanplane enters orbit around the Earth at an altitude of 400 km from its surface and becomes its satellite.

The driver-astronaut with the help of the electric rocket engine 6 brings the Titanplane into orbit of the international space station ISS. Controlling the Titanplane, he brings it to a distance of 3 km from the station and begins to move on a parallel course for rapprochement. The driver-astronaut, who is in the cabin of the TLC-1, undocks the capsule using docking unit 14. He switches on the chemical rocket engine of capsule 9. After that, TLC-1 departs from the Titanplane and is sent to connect with the berth N1 of the ISS. The takeoff and landing capsule TLC-1 docks with the ISS. The astronaut-driver, who is in the cabin of TLC-2 capsule, undocks the capsule from the space train using docking unit 15 (Fig. 4.1). With the help of a chemical rocket engine 10, it directs the TLC-2 to the berth N2 of the ISS.

After the convergence and docking of TLC-1 and TLC-2, the astronauts pass to the ISS. This ends the first regular flight of the Titanplane. To carry out the next regular flight of the Titanplane, it is necessary to refuel the electric rocket engine with a working substance—liquid hydrogen and refuel the fuel tanks and oxidizer of the chemical rocket engines of the capsules TLC-1 and TLC-2. The technology of such refueling was developed by the author during the design of the moonplane—a spacecraft for the regular delivery of astronauts to the Moon [9].

4.3 Structure of the Titanplane

The space train for carrying out the regular flights of astronauts towards Saturn's satellite Titan is shown on Fig. 4.1 where:

1. Space locomotive.
2. Emitter of heat into outer space.
3. The first tank—container with the working substance—liquid hydrogen.
4. The second tank—container with the working substance—liquid hydrogen.
5. Docking unit.
6. Superconducting electric rocket engine SERM-3K.
7. The first takeoff and landing capsule TLC-1.
8. The second takeoff and landing capsule TLC-2.
9. Chemical rocket engine for propulsion TLC-1.
10. Chemical rocket engine for propulsion TLC-2.
11. Shaft with nozzle for rotation of takeoff and landing capsules.
12. Superconducting bearing.
13. Docking unit of the tank-container.
14. The docking unit of the first takeoff and landing capsule.
15. The docking unit of the second takeoff and landing capsule.
16. Nozzle for the installation of takeoff and landing capsules.

4.4 Space Locomotive of the Titanplane

In the project for the movement of the Titanplane, a new version of the space locomotive was developed. At the same time, the basic scheme, which was proposed in 2014, is kept the same as in the space train for expeditions to the planet Jupiter [4]. The space locomotive has a cylindrical hull 2 with a conical nozzle 1. In the nose there is a complex of the main electric rocket engine 17, which is located along the longitudinal axis at the very end of the rocket.

A detailed description of the superconducting electric rocket engine SERM-3K is given below. The working substance of the electric motor is hydrogen, which in a liquid state is in the cryogenic tank 3. To increase the electrical conductivity of hydrogen gas, which in the working chamber of the electric motor is in a plasma state, at a temperature of 2500 °C, it is necessary to add 0.1% of cesium gas. With the help of cryogenic pipeline 19 liquid hydrogen is fed into the bow part of the locomotive where the on-board power plant 10 is located. Inside the cylindrical cavity of tank 16 along the axis of the locomotive is located the gas-phase nuclear reactor 9.

Fig. 4.6 Space locomotive of the Titanplane

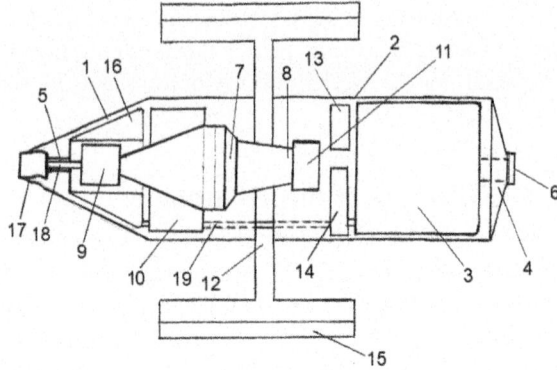

The cruiser electrorocket engine of locomotive 17 is mounted along its axis using a hollow cylinder 18. Outside the cylinder 18 is a pipeline and a heater system 5 are located, by which hydrogen gas is heated to temperature of 2600 °C and fed into the working chamber of the electric motor.

The presence of cryogenic tank 3 allows the locomotive to make short autonomous flights in outer space. For long-term flights, when the locomotive drives the space train, it is connected by a docking unit and a cryogenic pipeline to the container tank—the main carrier of the working substance. In the bow and middle part of the locomotive along its horizontal axis there is an on-board power installation, the layout of which is shown in Fig. 4.6.

The main elements of the thermal and electrical part of the on-board power plant of the Titanplane: cryogenic tank 3, docking unit 6, gas-phase nuclear reactor 9, MGD generator 10, superconducting generator 8, gas turbine 7 and compressor 11, sitting on the same shaft with turbogenerator 8. Part of the installation is a radiator 15 that provides heat transfer to outer space by heat radiation. Radiator 15 has a cylindrical shape. It is extended from the locomotive body by means of traction systems 12. The on-board power plant also includes a frequency converter 14, which connects the turbogenerator 8 to the MHD alternator 10 using an electrical cable. In the electrical compartment of the locomotive, a battery 13 is placed, which serves to start and provide warranty power to the power plant.

The on-board computer center carries out all operations during the operation of the unit and ensures its protection in emergency situations. A detailed description of the operation of the complex, with the help of which the electric propulsion of the Titanoplane is carried out, is given in [2].

4.5 Superconducting Electric Rocket Engine SERM-3K

To create the thrust force of the space locomotive, driving the space train to Titan, in the project an electric rocket engine of a new design was developed In the work [2] it was shown that the development and creation of a perfect electric rocket engine is the most important scientific and technical task in human exploration of outer space. As a result, he came to the conclusion that for orbital flights in the solar system it is advisable to use a magnetoplasma electric rocket engine. Magnetoplasma electric rocket motor belongs to the class of magnetohydrodynamic motors. MHD motor is an electromechanical device in which the supplied electrical energy under the action of a magnetic field is directly converted into mechanical energy of working substance (conductor in plasma state) movement.

The developed magnetoplasma engine has a working chamber into which a gaseous working medium is fed. Inside the working chamber, electrodes are installed to which voltage is supplied from the on-board source of electricity. Under the influence of an electric field, the working medium is ionized, and under the influence of the flow of electric current, it is heated. The working body goes into a state of plasma. Outside the working chamber, a superconducting magnetic system is installed, creating a transverse magnetic field in the space between the electrodes. When the current between the electrodes and the transverse magnetic field interacts, an Ampere force arises, which throws the working substance out of the nozzle of the working chamber, creating the thrust of the electric rocket motor. In the developed project, an attempt was made to create a superconducting magnetoplasma rocket engine with optimal parameters.

To carry out a comprehensive calculation of the electric motor, computer programs for the electrical, mechanical and thermal calculation of superconducting electric motors, as well as optimization programs for individual units and the engine, were used. The optimization criterion was to obtain the minimum weight and maximum service life of an electric rocket engine. As a result of these studies, a new design solution for an electric rocket engine was found. It became necessary to develop an optimal configuration of superconducting electromagnets that create the magnetic field of the electric motor. It turned out that the best performance is possessed by a magnetic system consisting of two counter-switched flat coils, the current in which is directed parallel to the longitudinal axis of the electric motor. In a magnetic system of this configuration, an intense magnetic field is formed not only between two electromagnets, but also on two sides outside the electromagnets. Therefore, to increase the magnitude of the magnetic field, it is in this space that the working chamber of the electric motor, which has a rectangular shape, should be installed. The optimal technical solution here is the installation of not one, but three working chambers of the electric motor.

The design of the electric rocket engine is shown in Fig. 4.7 (longitudinal section) and in Fig. 4.8 (cross-section).

The SERM-3K electric rocket engine has three working chambers. The first internal working chamber 1 is located along the longitudinal axis of the engine.

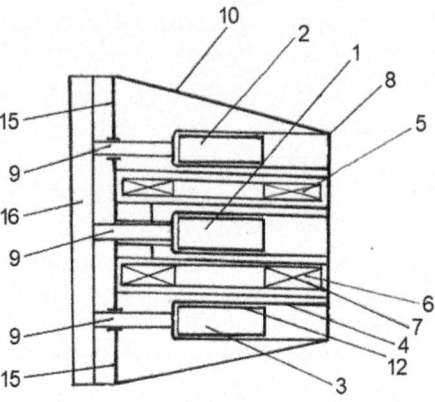

Fig. 4.7 Electric rocket engine (longitudinal section)

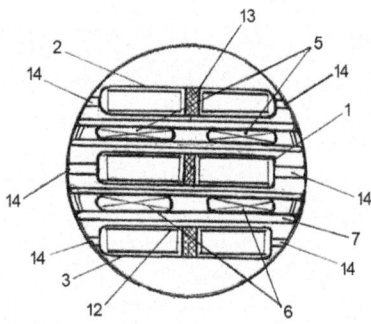

Fig. 4.8 Electric rocket engine (cross-section)

The second upper outer working chamber 2 and the third lower outer chamber 3 are arranged around the first chamber and the magnetic system. Each of the working chambers 1, 2 and 3 have a rectangular nozzle 4 and electrodes 12, which are made of tungsten. The electrodes are separated from the surface of the working chamber by insulating gaskets.

In each working chamber, the electrodes 12 are separated by insulating gaskets 13 that are made of high-temperature ceramics. Due to this, the flow of the working substance ejected from each nozzle is divided into two parts. To create a magnetic field in the working chambers 1, 2, and 3, the engine has a superconducting system, which is created by two counter-switched flat coils 5 and 6, which wind magnesium-boron from a high-temperature superconductor.

The coils of the superconducting magnetic system are placed in cryostat 7, which has a rectangular shape and is made of carbon fiber. The Cryostat has an internal cavity to accommodate the first working chamber 1.

All elements of the electric motor structure are connected to each other by means of an external conical housing 10 having a mounting ring 8 on the end side. The working chambers of the electric motor are attached to the housing by means of a

4.5 Superconducting Electric Rocket Engine SERM-3K

holder 14. The working substance is supplied to the working chambers 1, 2, and 3 by means of gas pipelines 9.

The attachment of the cryostat 7 to the outer cone 10 is carried out by means of a thermal bridge system 15, which provides a minimum heat flow. In the end part of the electric motor there is a shell cylindrical unit 16, where a system for supplying the working fluid 16 to the working chambers, consisting of a dispenser and a solenoid valve, as well as a system of electrical connections for supplying voltage to the electrodes 12 in the working chambers of the motor, is installed. The system of supplying voltage to the electrodes allows, by changing the direction of the current between the electrodes, to change the direction of the tractive force in the right and left parts of the working chamber and thus to exist the maneuver of the space train.

The calculation of the SERM electric rocket engine was carried out on a mathematical model, which allows to optimize its basic parameters. When designing the electric motor, an analysis of options with working chambers of various geometric configurations was performed.

The results of the calculation of the electric motor are shown in Tables 4.1 and 4.2.

Table 4.1 Design parameters of the superconducting electrorocket engine SERM-3K (1)

Technical parameter	Internal work camera	External work camera
Tractive force	1000 N	500 N
Power	10,000 kW	5000 kW
Voltage	1600 V	800 V
Current	6250 A	6250 A
Magnetic induction	2.6 T	1.3 T
Specific impulse	11,500 s	11,500 s
Efficiency	0.96	0.95

Table 4.2 Design parameters of the superconducting electrorocket engine SERM-3K (2)

Technical parameter	Value
Engine weight	1800 kg
Cone inlet diameter	0.85 m
Output diameter of the cone	1.1 m
Working channel length	1.2 m
Working channel height	0.1 m

4.6 On-Board Power Plant of the Titanplane

The scheme of the on-board power plant of a space train for flights to distant planets of the solar system described in detail in [9] and is shown in Fig. 4.9, were

1. Gas-phase nuclear reactor.
2. Superconducting MGD alternator.
3. Gas turbine.
4. Superconducting turbogenerator.
5. Compressor.
6. Radiation emitter.
7. Gas circulation circuit.

In this project, calculations of all elements of the on-board power plant shown in Fig. 4.9 have been performed. A description of the design of all elements of the on-board power plant is given in [9]. The design of a gas-phase nuclear reactor, a superconducting MHD generator, a gas turbine and a superconducting turbogenerator are given.

Reference [9] also provides an analysis of the electric propulsion system of the Titanplane. The results of the calculation of all elements of the on-board power plant and the electric propulsion system of the Titanplane are given in Tables 4.3, 4.4, and 4.5.

The results of the calculation of the masses of the locomotive elements are shown in Table 4.5

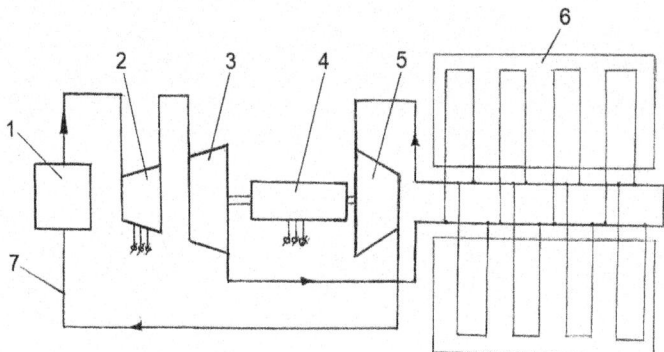

Fig. 4.9 Power plant of a space train

4.6 On-Board Power Plant of the Titanplane

Table 4.3 MHD generator parameters

Parameter	Value
Cone inlet diameter	0.8 m
Cone outlet diameter	1.1 m
Working channel length	1.5 m
Working channel height	0.08 m
Pole pitch of the winding	0.34 m
Input speed of the working substance	300 m/s
Output speed of the working substance	100 m/s
Voltage	4750 V
Current	3000 A
Current frequency	200 Hz
Magnetic induction	2.5 T
Power factor	0.45
Efficiency	0.82
Power	10,000 kW

Table 4.4 Superconducting turbogenerator parameters

Parameter	Value
Power	10,000 kW
Voltage	4750 V
Current	2700 A
Power factor	0.5
Current frequency	200 Hz
Rotation frequency	12,000 rev/min
Magnetic induction	2.8 T
Efficiency	0.98

Table 4.5 Design parameters of the locomotive elements

Parameter	Value (ton)
Locomotive	26
Nuclear reactor	15
MGD generator	2.5
Gas turbine	0.8
Turbogenerator	0.8
Accumulator	0.8
Tank with liquid hydrogen	2.5
Frequency converter	0.6
Case	2.5

4.7 The Takeoff and Landing Capsule (TLC) of the Titanplane

The capsule is designed to move 5 astronauts from Earth orbit to Titan's orbit. The design of the TLC is shown in Fig. 4.10.
  Here

1. The chemical hydrogen–oxygen rocket engine with thrust of 8 T and specific impulse of 400 s.
2. The bottom of the rocket body of the capsule.
3. Tank with liquid oxygen.
4. Tank with liquid hydrogen.
5. Chemical rocket engine for maneuver.
6. Cryostat tank with liquid hydrogen.
7. Expedition crew cabin.
8. Superconducting solenoid to create the magnetic field of protection against the flow of charged particles.
9. Docking unit for connection to the superconductor bearing shaft.
10. Tunnel and gateway to exit the rocket.
11. Parachute system for landing the capsule on the surface of Titan.
12. Instrumental container.
13. Landing tripod power ring.
14. Landing post damper.
15. Shoe boarding rack.
16. External case of the TLC.

  TLC mass on the Earth—26 tons. TLC length—14 m. External diameter—TLC 6 m.

Fig. 4.10 Takeoff and landing capsule

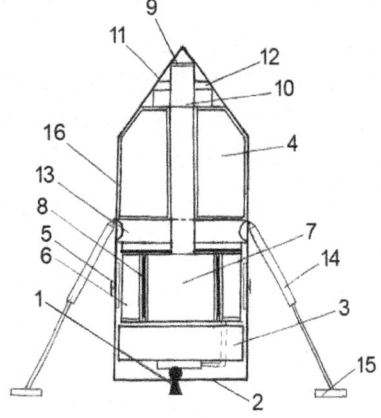

4.8 Conclusion

Work has been carried out to create the Titanplane—a space train for the regular delivery of astronauts to the surface of Titan.

The design of the Titanplane is developed, which is formed in the Earth's orbit by docking individual modules delivered from the Earth's surface by existing launch vehicles.

The movement of the Titanplane from the Earth's orbit to the orbit of Titan is carried out using a magnetoplasma electric rocket engine of a new design. The electric rocket engine has three working chambers in which the working substance is ionized using an electric field. The magnetic field in the working chambers is created by a superconducting excitation winding, which consists of two counter-switched on rectangular electromagnets with a current directed along the longitudinal axis of the engine.

A design development of an on-board source for powering an electric rocket engine was carried out, consisting of a gas-phase nuclear reactor, a superconducting MHD alternator and a superconducting turbogenerator with a capacity of 20 MW, a frequency of 200 Hz, which are located in the locomotive of the Titanplane.

The new design solution of the Titanplane with two takeoff and landing capsules will allow regular delivery to Titan of 10 astronauts in cabins that have a system of artificial gravitation and magnetic protection against cosmic radiation using a superconducting solenoid. At the same time, the flight time between orbits will be 65 days.

References

1. A. Rubinraut, The expedition to Saturn and creation of a space base on Titan. Int. J. Emerg. Technol. Adv. Eng. **6**(3), 1–11 (2016)
2. I. Bolvashenkov, J. Kammermann, A. Rubinraut et al., *Vehicle Electrification: On Water, in Air, and Space* (Springer, Switzerland, 2022)
3. S. Battersby, Titan's complex and strange world revealed. New Sci. **29**(10), 2004 (2004)
4. A. Rubinraut, The expedition to Jupiter. Int. J. Emerg. Technol. Adv. Eng. **6**(1), 283–293 (2016)
5. Iglus auf dem Mars. Space **4**, 80 (2017) (in German)
6. Mars Rover 920170. Space **4**, 71
7. A.E. Roy, *Orbital Motion* (CRC Press, Boca Raton, 2020)
8. D. Leisner, Der Späher im Reich des Ringplaneten. Geo Kompakt 21-12/09 (2009) (in German)
9. A. Rubinraut, Moonplane—a spacecraft of regular delivery of astronauts onto the Moon. Adv. Aerosp. Sci. Technol. **4**(3), 43–56 (2019). https://doi.org/10.4236/aast.2019.43004

The manufacturer's authorised representative in the EU is Springer Nature Customer Service Centre GmbH, Europaplatz 3, 69115 Heidelberg, Germany. If you have any concerns regarding our products, please contact ProductSafety@springernature.com

Printed and bound by CPI Group (UK) Ltd, Croydon, CR0 4YY
26/03/2026
02078940-0011